DATE DUE

DEMCO 38-296

PRIVILEGED
HANDS

PRIVILEGED HANDS
A Scientific Life

GEERAT VERMEIJ
Department of Geology
University of California at Davis

W. H. FREEMAN AND COMPANY
NEW YORK

	)dder
	mino Maass
Cover and Text Designer:	Diana Blume
Cover Photographer:	Gerry Gropp
Illustration Coordinator:	Susan Wein
Illustrator:	Roberto Osti
Production Coordinator:	Sheila E. Anderson
Composition:	Digitype
Manufacturing:	R. R. Donnelley & Sons Company

Library of Congress Cataloging-in-Publication Data

Vermeij, Geerat. J., 1946–
    Privileged hands : a scientific life / Geerat Vermeij.
        p. cm
    Includes index
    ISBN 0-7167-2954-7 (hardcover)
    1. Vermeij, Geerat J., 1946– ,.   2. Biologists–United States–
–Bibliography: 3. Blind biologists—United States—Biography.
I. Title
QH31.V39A3 1996
574′.092—dc 20
    [B]   96-21792
                                                              CIP

Printed in the United States of America

First printing, 1996

CONTENTS

FOREWORD

What would the blind lad get out of this trip? That was the question running through my mind as we drove the university vans across the New Jersey coastal plain from Princeton to Long Beach. We were on the first field trip for my undergraduate course in invertebrate paleontology and planned to collect organisms, particularly molluscs and their shells, for later study. My students and I would learn to decipher fossil records by collecting and examining all kinds of seashore shell life, both dead shells buried in sediments and sample animals from living communities. I had met Gary Vermeij only recently in my colleague Newell Brown's office. Almost immediately, I recognized his great promise. Still, our field excursion would be a test.

We drove across the lagoon with its fringes of tidal marsh and parked the vans. I am sure that Gary's nose and ears alerted him to the proximity of the ocean, hidden from our sight by the dunes. We trudged across the dunes, Gary lightly holding a classmate's elbow, and stopped in view of the sea.

We held a brief discussion of the natural setting, planned our activities, and scattered to make observations and collect shells, which seemed few and far between. Gary immediately felt his way down to the wet sand at the swash line, the line where the waves wash up on the beach, then back up the beach face to the line of flotsam and jetsam left by the last high tide. On his knees, hands buried in the little ridge of drying seaweed, gum wrappers, and plastic bottles, he set to work.

Ten minutes later, when the class regrouped, he was still busy. Raising a shell high over his head, he exclaimed, "I can't believe it! Tell me, is this shell pink?" We assured him that it was indeed, whereupon he added, "What the hell is it doing *here*? It's a *Tellina* and has no business being as far north as New Jersey!"

So much for my questions about Gary. His collection of shells that day was the best by far. His fingers knew the species, and their names were etched in his memory. Later that day, as we stood ankle-deep in the flooded tidal marsh, spading up a bit of the *Spartina* peat and the horse mussels nested within, Gary, fingering the bottom, discovered the line along which the horse mussels gave way to the blue mussel *Mytilus edulis*, which required more contact with seawater.

And so it went through the rest of the semester. Instead of floundering without sight, Gary had learned to live without it and in so doing developed lifelong goals and a maturity that far exceeded that of his contemporaries. Instead of his classmates assisting him, he would help them.

In this book, Gary reveals how he learned to live with his blindness. Many things helped. One was his recollection of sight and colors from before age four, when he lost his eyes. Another was a wonderful family. His parents, while struggling financially, respected intellectual values and provided their loving support for Gary's ambition, although it must have seemed a pipe dream. They were not too protective, and they allowed him to experience the world as widely as possible. His older brother Arie was very helpful throughout their childhood together.

Life in a Dutch boarding school provided an excellent education, although not without some social trauma. What we must read between the lines is the degree to which Gary's personal charm and precocity won him friends. These friendships later helped to open the vast resources of libraries and museum collections normally beyond the reach of the blind.

Several important events marked the direction of Gary's life. The first was probably the inspiration provided by his fourth-grade teacher, Mrs. Colberg, who sparked his interest in shells and showed him her own collection of shells from Florida. These shells were very different from the shells he knew on Dutch beaches and piqued his curiosity about what caused the differences. Other milestones include the immigration of Gary's family to New Jersey when he was nine; his first contact with Henry Coomans and William Old, two malacologists at the American Museum of Natural History; and his first subscription to *The Nautilus*, the journal that revealed to him the searching and dialectic nature of science.

Later, Gary's scholarship to Princeton University provided the springboard to his burgeoning academic career. Our mutual friend Egbert Leigh has characterized the university, from his own experience, as a place where "any half-bright student could get all the professorial attention he could handle." And Gary had just the right mix of qualities—curiosity and devotion, skills and independent thought—to qualify for such attention. I will never forget the occasion when Robert MacArthur was lecturing in biogeography and illustrating evolutionary convergence by showing us two bird skins from the university collection—one an American meadowlark, the other an African look-alike from a different family. Gary re-marked, "I don't find any resemblance. The beak and the feet are entirely different, and the very texture of the plumage sets them apart."

Not only has Gary surmounted his presumed handicap, he has learned to profit from it. Other, sighted naturalists have combined their deep love for nature with an intense curiosity, a lively imagination, and a keen brain. Gary would have become a fine naturalist if he could see with his eyes, but he sees with his hands, and this provides him with an awareness of details to which our reliance on vision has blinded us. The sighted who have collected shells across Earth's latitudes have consis-tently noted the drabness of cold-water molluscs compared with the bright coloration of molluscs found in tropical zones. Gary wonders further why the former feel soft and chalky and the latter hard and glossy.

Our eyes see mainly in two dimensions. Gary experiences form palpably in three dimensions, which provides him with a different and often advantageous perspective. Each scientist brings an individual perspective, of course, conditioned by his or her own genetics and experience, and that is what keeps sci-ence lively. But Gary's perspective differs from the norm by a quantum leap. Not only has he substituted tactile sensing for sight, but the very lack of trivial visual diversions in his life taught him the habit of serious mental reflection from child-hood.

Gary completed his Ph.D. at Yale University. While there, his relationships with other students and his interest in their studies of molluscan ecology and biogeography on a global scale opened to him the fourth dimension of time, and he be-came fascinated with fossil shells as gateways to the distant past. After graduate school, he embarked on a teaching career that started at the University of Maryland and continued at the

University of California at Davis. During his time at Davis, he won one of the coveted MacArthur fellowships.

Research and teaching of evolutionary biology are deeply dependent upon specimen collections, which are mostly housed in the major museums to which Gary acknowledges a major debt. In addition, his familiarity with the vast resources of literature, made accessible to him by readers, has led to his editorship of several important scientific journals. Equally vital to his work is the extensive travel that has allowed him to make comparative observations on the life of molluscs and their predators in various ecological and geographic settings. He is one of the few practicing scientists who is equally at home in ecology and evolution, as well as understanding their paleontological dimensions.

Is the appearance of shell-cracking crabs responsible for the evolutionary surge of molluscs during Mesozoic and Cenozoic time? It will be a pleasure to follow Gary's research. I hope that this singular man's experiences in pursuit of elusive and ever-tantalizing Nature will lead to more books that open insights into the nature of scientific enterprise.

Alfred G. Fischer
San Pedro, California
July, 1996

Chapter 1

THE FIRST QUESTION

A tray of shells stands before me in the bright December sunshine streaming in through the north window of our cozy rented house. I know that the previous day's haul, which we had collected by dredge while on board the *Munida* from a rich community of sponges and pen shells off the Otago Peninsula on the South Island of New Zealand, will yield surprises. It will prompt questions that I would not have known to ask, and it will steer me to new thoughts that only firsthand observation can provoke. I turn over each shell in my hands, employing fingertips and nails to scrutinize all the subtleties of shape and surface ornament. There is astonishing diversity here. A hand-sized star shell, its long hollow spines jutting from the flattened whorl-like sun rays, competes for space in the tray with hairy mussels, bearded ark shells, and huge fan-shaped pen shells plastered with colonial sea squirts and the long sinuous tubes of worm snails. A thick layer of sponge obliterates the delicate scaly ribs of a scallop. Thin slipper limpets tumble out of the openings of snail shells

1

in which hermit crabs had found a secondary home. Borers had attacked the large plain triton shells so thoroughly that the thin internal glaze of the shell is pockmarked with patches where the snail builder was compelled to repair the damage.

In my pastoral setting, with the insistent cries of lambs for their mothers drifting in from the grassy hills beyond, I have time to contemplate what I have seen in the field. How odd that all the cold-water shells from New Zealand are so thin. Not only the deep-water snails and clams brought up yesterday by the *Munida* are lightly built, so are the dogwhelks, mussels, turban snails, and even oysters from the nearby rocky shores. Certainly they contrast with the thick-shelled productions so characteristic of molluscs in Chile, Maine, and the Pacific Northwest. The rich diversity of surface sculpture also attracts my notice. Spines of the sort displayed by the star shell are common enough in tropical shells from comparable depths in the western Pacific but rarely adorn shells from cold regions. Do such architectural features reflect the heritage of a tropical ancestry? Perhaps, but several of the spines appear to have been broken and then regrown when the snail builder was alive, suggesting that they might have served a valuable function in the living animal.

In other ways, the snails and clams of Otago conform faithfully to expectation. Like cold-water shells everywhere, most reveal a coarse chalky texture, not the hard marblelike feel of warm-water shells. Why should this be so?

Thirty-seven years earlier and half a world away, I asked that same question. Quietly and without fanfare, perhaps even without any conscious thought on my part, that question transformed an ordinary day in fourth grade into a day when all the enjoyment I had always felt for natural objects crystallized into a burning curiosity.

That morning must have dawned like any other in the fall of 1956. As my brother Arie and I made our way from the rambling second-floor apartment on Salem Street to East Dover Elementary School, the sweet aroma of decaying leaves and the fermenting crab apples that lay trampled on the pavement was interrupted only briefly by car fumes. Within five minutes I was in my seat near the window in the front row, within easy reach of Caroline Colberg's large wooden desk. The first lesson dealt with the early American explorers. Their names and the places they conquered—Cortez in Mexico, Pizarro in Peru, Cartier and de Champlain in Canada—fired my imagination. I pictured thick jungles and rugged mountains, and the greed and courage of Spaniards slogging across Panama to gaze upon the Pacific. Mrs. Colberg enlivened the lesson with tales of her own youth in Panama at the time the great canal linking the Atlantic and Pacific oceans was nearing completion.

History had to make way for art. Although pictures drawn with pencil or pen held no meaning for me, Mrs. Colberg insisted that I participate. My assignment was to produce a catalogue of leaf shapes, each leaf carefully traced with a stylus on a sheet of Braille paper laid on a board to which window screening had been nailed. A sharp raised outline of the leaf would appear on the back side of the paper, to be carefully labeled. The simple ovate leaves of cherry, apple, and lilac posed no problem; maple and oak, with their lobes and toothed margins, challenged my untrained hand.

My habit of working quickly through assignments left me with oceans of free time. Oblivious to what the rest of the class was doing, I avidly lost myself in the pages of the newly acquired seven-volume Braille dictionary that stood on the long table at the front of the room. Ludicrously, it bore the title of Vest-Pocket Dictionary. For a time I wondered if the publishers had meant "Vast" instead.

But on this day there was something new to divert my attention. Always eager to decorate the wide windowsills of her sunny classroom, Mrs. Colberg had brought to school a few shells and pieces of coral that she had acquired on her frequent travels to the west coast of Florida. My strategic location next to the display afforded ample opportunity to sneak a quick look.

I was prepared to like what I saw. Back in the Netherlands, I had already grown fond of shells. A successful day at the beach meant a good haul of cockles, wedge shells, and razor clams. Cockles, their valves neatly adorned with a fan of ribs ending at the crisply saw-toothed margin, contrasted with the plainer clams in which the uneven riblets ran parallel to the smooth edge. However I might have liked them, these chalky productions paled in comparison to the elegant Florida shells. Mrs. Colberg's finds felt as if they had been crafted by a sculptor with an eye for regularity and intricate detail. The ribs on the cockle were crisper, more prominent, and adorned with a flourish of little overlapping scales. The shell interiors were not dull, but smoother and more polished than I had ever imagined possible, so that fingertips glided over their surfaces as they would on glass. There were snails as well, often with the most unlikely shapes. How could one explain a shell as odd as the lightning whelk, with a spiral crown of knobs at one end and a drawn-out spout at the other? Why was its interior so stunningly sculptured with smooth, evenly spaced ribs that spiraled away beyond the reach of my fingers?

Daily the display grew, as my classmates, many of whose fathers had fought in the Pacific during the Second World War, brought shells from home. A shell from the Philippines I now know as *Tectarius cornatus* presented a perfectly conical shape, its whole surface evenly sprinkled with sharp glossy beads set in tight spirals. Cowries challenged my notions of what was possible in the realm of nature. The exterior was so implausibly polished and so evenly domed that I believed someone had applied an especially thick coat of varnish to the shell so that all of the sculpture was obliterated. Not until much later did I come to understand that the polish is natural. In the living animal, the shell is enveloped by a retractable flap of mantle tissue, the inner surface cells of which secrete the glaze. Most intriguing of all was the large heavy Florida helmet shell that Trina Mendon brought. Beautifully rounded knobs separated by thin finely beaded ribs adorned the domed roof. The flat base consisted of a polished shield, in the middle of which the long narrow opening of the shell gaped like a deep rib-lined canyon. The precision of the ribbing and the pleasing contrast between the glaze of the base and the coarser fine-grained stone texture of the dome produced for me an architectural splendor completely outside the realm of my previous tactile experience.

Mrs. Colberg told of the beaches on which one could casually gather these works of art. I daydreamed of such places. They would bear exotic names, and gentle waves of warm water would reach up from below, depositing shells in which corrosion never compromised textural complexity. I wondered why the cold-water shells of the Dutch beaches were so chalky and plain, whereas tropical creations were ornate and polished. My fourth-grade teacher had not only given my hands an unforgettable esthetic treat, but she aroused in me a lasting curiosity about things unknown. None of it was in the books; there was no expensive conspiracy to teach science, no contrived lesson plan painstakingly conceived by distant experts. Instead, Mrs. Colberg captured the essence of her task. She created an opportunity, a freedom for someone to observe, an encouragement to wonder, and in the end a permissive environment in which to ask a genuine scientific question.

Once awakened, my curiosity knew no bounds. I wanted shells of my own, and I longed to know their names and the habits of the animals that built them. By February 1957, a few cigar boxes held the beginnings of my collection. Using wooden crates scavenged from local grocery stores, my father built a sturdy cabinet of six open compartments, each lined with vinyl sheeting to hide the rough wood beneath, in which I could store my shells and books.

Arie enthusiastically took up reading books out loud. While he read, I transcribed place names, descriptions of shores, and facts about sea life. With this routine of what we christened "sea school" firmly established, Arie solidified his role as teacher by illustrating my notes with carefully drawn Braille illustrations. He filled page after page with faithful renderings of seaweeds, sponges, worms, jellyfish, shells, sea stars, crabs, fishes, and birds, all accurately labeled in Braille. We read through atlases to learn about the world's coastlines, and inevitably Arie turned his talents and geographical curiosity to drafting detailed Braille maps. Features of interest were indicated with letters and numerals, for which he provided a key on an accompanying page. With this extraordinary dedication and the clever use of minimal household technology—plywood, window screening, and a stylus—Arie and the rest of my family effectively broke the information barrier by making available the full richness of the print media to me. Even at this most elementary level, the existing Braille books in libraries were wholly inadequate. I cannot remember a single book about shells in any Braille library. Braille publishers of the day viewed raised illustrations as unintelligible and produced only maps that were so uncluttered that all but the largest towns and physical features of the land were left out.

I proclaimed to anyone who cared to listen that I would be a conchologist. What's that, people would ask. I barely knew myself, of course, but books told me that people who study and collect shells are called conchologists. My parents, who strongly encouraged their two sons to develop serious hobbies, did everything they could to deepen my commitment. If Arie could collect stamps, and in the process acquire a formidable grasp of geography and world events, why shouldn't I devote my energies to shells? Mrs. Colberg, too, egged me on, even giving me a few of her prize specimens. When one of the fifth-grade classes visited the American Museum of Natural History in New York, they returned with a box for me, full of the most marvelous shells, each one glued to the bottom of the box and accompanied by a formal Latin name and the place from which it came. I now possessed my very own *Tectarius coronatus* from the Philippines. There was a *Strombus canarium,* smooth on the outside but even more so inside, from the mysterious Great Barrier Reef of Australia. Next to it sat a silver-lipped stromb, *Strombus lentiginosus,* intricately knobbed and corded externally and again thrillingly polished on the underside around its narrow aperture.

If Mrs. Colberg harbored doubts about my ambitions, she kept them to herself. My obsession must have looked like any boy's fanciful dream of becoming a fireman, baseball player, or spaceman. Sooner or later, this blind boy would settle on a career, to put it discreetly, more consistent with his limitations. Yet none of this was said. Instead, there was unanimous and unreserved encouragement.

If that autumn day in 1956 passed unnoticed for Mrs. Colberg and the rest of her class, for me it was like no other. On that day, a wonderful teacher set the course for one man's life.

Chapter 2

A GLOW OF YELLOW

The pebbles sparkled along the narrow gravel path beside the canal. I paid little attention to my mother's repeated requests that I walk upright as I gazed down at the dance of sparkles passing beneath my feet. There would be more than enough time to straighten up once we reached the dull paving bricks of the streets closer to home. Meanwhile, the changing display of shadows and reflections offered a welcome diversion from the relentless pain in my eyes.

The pain had been with me since that warm sunny day in September 1946 when I was born at Sappemeer, a farming community built on the reclaimed fens in the northeastern Dutch province of Groningen. Doctors soon concluded that I was afflicted with an unusual childhood form of glaucoma. Swelling enlarged my eyes, and medicines relieved the pain only temporarily.

Most of the next three and a half years of my life were spent in hospitals, first in Groningen and later in Utrecht. Though fragmentary, my memories of this period remain vivid.

A kindly nurse occasionally allowed me to tear newspaper and to crinkle aluminum foil in her room during free hours. Drawers beneath my bed held a few toys. I remember liking the food.

Once, when my mother brought me to the hospital at Utrecht, I cried and cried as she disappeared slowly from my view. A nurse scooped me up in her arms and whisked me to bed before I could fully take in our parting.

A dreadful routine developed at Utrecht. Operations on my eyes were scheduled every Wednesday and Saturday. As I was being wheeled into the operating theater, I could already smell the faint traces of chloroform that warned of things to come. My head would be firmly fixed in place while the surgeons tended to my eyes. I struggled to keep my promises not to cry, knowing that I would be given the empty metal spools on which bandages had been wound if I succeeded. Later, in bed, I would make elaborate constructions by clamping the spools together. A few days of respite passed before the next operation. On Tuesdays and Fridays, my mother came to visit for a half hour. I insisted she read me the same nonsense verses over and over. When my father came on Sunday, the verses would be recited yet again.

Only when my own healthy daughter was born in 1981 did I begin to comprehend fully how trying it must have been for my parents to maintain a sense of normality at home in the face of constant uncertainty and the emotionally draining trips to and from the hospital. There was not much money, and the rest of the family offered little support, financial or moral. Yet they persevered, and very well at that. Mine was only the latest chapter in a long story of frustration and hardship.

My father, Johannes Lodewijk Vermeij, was born the eldest of three children in 1915 at Schiedam. His mother, Geertruida Overeinder, died of a mysterious ailment at age fifty-four in 1942. His father, Arie Pieter (whom we called Opa Vermeij), was a stubborn, domineering man who did his best to thwart my father's wishes to break the family tradition of working at respectable office jobs. Opa's house on the Krugerlaan in Gouda reeked of the cigars Opa smoked until his death of lung cancer in 1961. Against Opa's wishes, my father pursued academic training in horticulture and arboriculture at Boskoop, with the intention of entering the nursery business so that he could indulge his love of plants. Military service, however, intervened. The German invasion of the Netherlands in May 1940 found my father in the armed forces, and it might have cost him his life had his platoon not been diverted at the last moment from duty at the Grebbeberg, the scene of a troop massacre.

Rumblings of war were already in the air when my father met Aaltje Hindrika Engelina Smith on holiday in Luxembourg in 1937. Four years later, they were married in The Hague.

My mother was born in 1912 to a prosperous family of gentlemen farmers at Sappemeer. The farm, which has long since disappeared, raised wheat, potatoes, and vegetables. When her parents separated in 1921, my mother and her brother Seitse stayed with their mother, Nantje Burema, while their father, Geerat Jacobus Smith, moved to Hillegersberg near Rotterdam with a German woman we knew as TaNte Mila. Seitse, who suffered from a congenital heart condition, received much of Oma Burema's attentions while my mother fended for herself. After completing secondary school, my mother worked in a pharmacy, preparing medicines from diverse dried herbs for hours on end. Meanwhile, Opa Smith built up a thriving export business in vegetables and traveled widely throughout Europe. Whenever he and TaNte Mila came visiting, he would arrive in a huge American car, the only such vehicle owned by anyone in my family.

During the war years, my parents lived in The Hague. My father worked in the food distribution administration and kept a low profile lest the German occupiers spirit him away as a slave laborer in the factories of the Third Reich. The most trying times came toward the end of the war. It was during the winter famine of 1945 that Arie Pieter was born. In desperation, my father journeyed by bicycle to Sappemeer to fetch a supply of beans, potatoes, and rye bread from Seitse's farm.

After liberation in 1945, my family moved to Sappemeer, where my father worked for a time on Seitse's estate and later in the office of a bakery at nearby Martenshoek. Neither job being very suitable, plans were made to return to Gouda, where my father eventually found employment at the same candle-making company where Opa Vermeij had long been a manager. While I remained in hospital at Groningen, the rest of the family moved to a modest flat at Oosthaven 63 in early 1947, in the old part of Gouda. Two years later, we moved again, this time to a cozy five-room flat at Koningin Wilhelminaweg 371, in a newer part of town known as the Korte Akkeren.

Arie and I each had our own tiny long room. One end opened onto an L-shaped hallway, the other via a glass door to a narrow balcony overlooking a meadow and a canal. One could hear sheep bleating and barges chugging lazily along the waterway. All the floors at home were covered with prickly coconut-fiber mats, unsuitable for racing cars but ideal for scraping bare knees. On cold winter evenings, we crawled into bed with jugs of hot water at our feet. I would often lie awake, listening to the deep reverberations of distant thunder in the summer or to the groaning of the concrete walls as fierce North Sea storms lashed the land and strained the dikes in winter.

Despite my poor vision, which enabled me only to distinguish colors and some vague images, I found the outside world beautiful to look at. Bright colors especially pleased me. I wondered why the sun looked yellow during the day but turned orange as it set. On the walk from the railway station to the hospital at Utrecht, we would pause at a particularly bright orange door.

It was becoming obvious that the measures being taken to relieve the swelling and stem the pain were not working. My parents now faced decisions with far-reaching consequences. Should further attempts be made to salvage what little of my vision remained, or should the risk of brain damage be eliminated once and for all by having my eyes removed? Professor Henricus J. M. Weve, the chief surgeon and the country's foremost ophthalmologist, recommended the latter course. In his early sixties, Weve was a gifted surgeon who would perform more than five thousand operations in his career. The youngest of Queen Juliana's four daughters, Princess Marijke (now known as Princess Christina), was also under Weve's care with an ailment similar to mine, probably the result of the German measles from which both the queen and my mother suffered during pregnancy. Artificial eyes, Weve explained to my parents, would eliminate the pain as well as the threat of brain damage and, most important, would create a normal appearance. Rather than draw attention to my blindness, the new eyes would enable others to see me as a normal human being. Wisely, my parents accepted Weve's advice, and authorized the removal of both eyes.

The final operation took place in June 1950. On the operating table I saw a glow of yellow. And then there was nothing.

Professor Weve, who enjoyed a well-deserved reputation of maintaining an excellent rapport with his patients, strongly advised my parents to treat me as they did Arie. Pity, either for themselves or for me, was intolerable, he said. Regard blindness as a nuisance that can be overcome, not as an unmitigating tragedy or debilitating affliction, Weve counseled. Accord him no special favors, lavish attention as much on him as on his brother, and let him enjoy the gifts of the world as others do.

After a short recovery period in the hospital, it was finally time to come home. I headed straight for my favorite closet where all the familiar toys were kept. The colors of the pencils were now gone, but my father had notched each pencil in a distinctive pattern so that I might still know which color went with which pencil. He cut letters and numerals out of paper so that I could learn what Arie already knew.

If I had difficulty adjusting to blindness, the memory has faded. Almost immediately I discovered the value of echoes for telling me where I was. Sounds bouncing off obstructions provide cues about the size of a room, the position of a tree, the speed of a bicycle or car, the presence of a person, whether a door is open or closed, and much more. It wasn't that the remaining senses became more acute now that I was blind; I simply relied on them more. The information they conveyed now meant something, whereas previously I could afford to ignore it. My world was not black and hopeless. It sparkled as it did before, but now with sounds, odors, shapes, and textures.

A few weeks after the final operation, we traveled to Rotterdam for my first pair of artificial eyes. Technical advances during the war had made it possible to mold eyes of a natural appearance by brushing them with lines representing the tiny blood vessels and by painting them the appropriate color, blue in my case. Each eye was a curved plate, convex on the outside, concave on its inner face, fitting snugly into the eye socket and held firmly in place by the eyelids. My first few pairs were made of glass, but they predictably shattered. Soon the switch was made to Lucite, surely one of the more bizarrely coincidental trade names for a durable shatterproof plastic out of which eyes for the blind are made.

Indeed, it is all these sensations that together provide a vivid, if nonvisual, picture of the world around me. The three-dimensional architecture of a forest comes to life with the sound of every step, every breath of wind, every chirp of a cricket from the underbrush or song of a bird from high in the trees. All these sounds bounce off the innumerable obstructions of leaves, branches, and trunks from ground level to high above my head. Raindrops descend slowly and noisily as they collide with leaves. High overhead, the wind hisses through the upper reaches of oaks and beeches or moves through needled pine branches with a low muffled whistle. With so many manifestations of nature making sound and soaking it up, I feel a strong sense of close intimacy with nature, of being surrounded and enveloped by it. That perception is reinforced by the occasional rays of warm sun penetrating the tree crowns and by the rich fragrance of moist soil, lush foliage, and the sweet holly or honeysuckle that advertise for pollinating bees in spring. What the ears and the nose perceive from afar, the hands can sense at close range. There is a pleasing diversity of textures and shapes among leaves—the leathery, saw-toothed leaf of the oak, the woolly three-leaf of a red clover, the curled-up young frond of a fern—and the litter on the forest floor may yield pinecones, acorns and their scaly cups, triangular smooth beechnuts in their thorny husks, or just a piece of brittle bark. I might happen upon a ring of mushrooms, or an inviting carpet of soft moss, or the left-twining stems of honeysuckle. Dead leaves on the ground on a warm day are crisp and curled, and they lack the cool moist feel of living leaves whose main characteristic for the sighted observer is the green color I do not see.

How different is the feel of a lush summer meadow. The breeze blows unimpeded against my face, carrying with it the sweet smell of green grass or dry hay and quietly rustling the low plants as it sweeps by. The continuous babble of a meadowlark overhead disperses far and wide over the land without resistance from trees, or it may echo off a distant row of poplars. The sun warms the skin and brings out the sweet fragrance of a patch of chamomile or flowering white clover. I cannot appreciate the colors, the views of majestic clouds, the pastoral scenery of trees or church steeples in the distance, the cattle and sheep grazing quietly in a nearby meadow; but there is so much here to enjoy, so much to take in, such a richness of sensation, that I can hardly mourn the loss of sight.

Yet opinion polls almost unanimously portray blindness as the most feared of human afflictions. Sight is perceived as the means by which we gain the bulk of our information about one another and about our surroundings. Accordingly, educators have built curricula almost entirely on a foundation of visual learning. For this reason, blind people are widely regarded as being incapable of learning or of interacting fully with others. Skeptics despair that blind people cannot see facial expressions, cannot witness a baby's first tentative steps, cannot respond to a smile, cannot see how others behave. Without such quintessentially visual experiences, the argument goes, the blind are denied a basic dimension of what it means to be human. Naively, those who fear or loathe blindness hope that feeling people's faces will somehow provide the cues that the eyes would otherwise perceive, and that such an act will enable the blind to know a person better. It is inconceivable to them that knowing what someone looks like simply does not matter to me.

What these sentimentalists forget is that the face is only part of the whole person. The voice—its quality, its intonation, the use of language—is unique and every bit as informative as the face. I can detect surprise, disgust, pleasure, boredom, dishonesty, thoughtfulness, and a hundred other states of mind from the voice. I may not use appearance as a criterion for attractiveness, but I certainly rely on a person's voice and on my senses of touch and smell. Even physical movements provide a fund of information. A quick, steady gait leaves me with a very different impression from a slow, uncertain shuffle. It seems to me that many people pay better attention to their visual appearance than they do to their voice and especially their physical demeanor, which makes reliance on visual cues less credible. Even a very small child learns quickly to assess people and their motives through nonvisual cues. The loss of vision does not rob the blind of their ability to learn about the world.

Does this mean that blind people can learn everything there is to know about their environment without help? Of course not. One of the great benefits of vision is that it enables people to perceive the world at a distance and on a large spatial scale. My understanding of the things around me was immeasurably enriched by my parents' continuous commentary about the passing scenery as we walked, bicycled, or traveled by train. Clouds, colors, cows, cars—everything mundane or out of the ordinary—were mentioned, described, and, when possible, demonstrated. In this way, I could integrate the world as people saw it into my own sensory world and let my imagination fill in the blanks.

Colors remained very much in my imagination. I cherished their memory and played endlessly with them in my mind. Each letter of the alphabet, each numeral, and many place names came to be associated with a particular color. The letters *b* and *v* were red, *m* and *s* were white, *d* and *j* and 2 and *Gouda* were yellow. Colors representing the letters would then be blended to form words. There were very few pure green words—*duif* (yellow *d* and *u*, blue *i* and *f*, for example—but lots of orange ones.

My parents missed no opportunity to treat me as a fully capable and responsible member of the family. An old Dutch proverb summed up their philosophy: "Ik kan niet, zei de dwaas, en daarom kon hij niet" (I cannot, said the fool, and that is why he could not). Nothing was out of reach; nothing Arie was allowed to do was off limits to me. I could go fishing as he could. If Arie could be asked to fetch the bread and cheese we brought for a picnic in the polder, so could I. ("Be careful not to fall into the ditch," my mother would say.) Usually I was very careful, but on one occasion when I was bringing Arie a pear I tripped and found myself with my legs dangling into the muddy water of one of the innumerable ditches that drains the land. My parents took such little mishaps in stride. How could one learn about the world if one was sheltered from it? They would have none of this, and neither would I.

Gouda, situated at the branching of the rivers Gouwe and Hollandse IJssel, is justly famous for its excellent cheese, which was traditionally made on farms in the surrounding polders. Meadows, ditches, and rows of poplars and alders stretch in all directions around the town, toward Waddinxveen and Boskoop to the north, Oudewater and into the Krimpener Waard to the south across the IJssel. We never tired of the lush pastoral waterlogged land. With Arie on the back of my father's bicycle and me on my mother's, we would set off eastward on the twisty dike along the IJssel and then descend to Het Laantje, one of the innumerable gravel roads winding through the countryside. In February we eagerly awaited the

first yellow flowers of figwort and the tiny daisies whose little
heads barely poked up above the damp clay. Later there would
be taller and showier flowers—marsh marigolds, poppies, but-
tercups, cuckoo flowers, fragrant chamomile, and red and
white clovers—interspersed with the odd stinging nettle. The
sweet aroma of grass blended with the pervasively sharp scent
of cow manure to produce an unforgettable olfactory signature
of this low rich land. If the wind was just right on a Sunday
morning, one could hear the faint tolling of bells from a distant village steeple, echoing and dancing with the breeze as it
must have done for centuries.

Weather permitting, we would take longer rides. The beach
at Scheveningen lay some thirty kilometers west of Gouda, a
good distance by bicycle but still just feasible. The sea announced itself far inland. First came the dunes—steep, sandy,
very still, covered and held in place by grasses and dry shrubs.
At their summit the vegetation thinned, and the breeze carried
with it the far-off roar of the surf and the smell of wet salty
sand from below. A wide expanse of sand still separated us
from the tideline where I wanted to be. Building sand castles
was not for me. I concentrated on the shells washed up at the
high-tide mark. I would return home with a sack full of sand-
caked clam valves ready to be sorted the next day.

At home, I developed a lasting taste for classical music. After my parents gave me a subscription to the Braille edition of
the weekly radio schedule, I came to know the names of all the
major composers and the instruments for which they wrote.
Music filled a void of loneliness with a deep feeling of warmth
and contentment. I cherish its complexity and its varied
acoustical textures—a warm Beethoven sonata for cello and
piano, the bright tinkling of a harpsichord contrasting with the
reedy bassoon, a sharp trumpet clashing with the more sub-
dued and refined violin, a great organ hurling the giant chords

of a Bach fugue at the stone walls of an ancient cathedral—
oh, how I might have liked to create such monuments! To this
day, I associate Beethoven's Sixth Symphony with a quiet
sunny room filled with the scent of honeysuckle. My musical
passion, however, remains entirely passive. I lack any talent for
performance, and composition is simply unimaginable.

My mother conformed perfectly to the ideal of a scrupulously clean and tidy Dutch household. She waged a constant
and almost entirely successful war against dust and clutter.
Seventeenth-century Dutch painters would have found the interior familiar. Thick fringed woolen rugs, their soft expanses
interrupted here and there by fine pieces of silver and brass,
graced the top of every table and cupboard. At meal time, the
dining-table rug was carefully replaced with a linen tablecloth,
which was decorated with my mother's own meticulous em-
broidery. Each place would be set according to strict custom:
knife on the right, its blade resting on a little glass support so
as not to sully the cloth; fork on the left; dessert spoon or fork
above the plate. Food was placed in the center of the table in
serving dishes. A cloth napkin neatly rolled in individually distinctive rings marked the left-hand border of each place setting. At the end of the meal, all the dishes and utensils were
cleared away promptly to be hand-washed, dried, and properly
stowed before the rug reappeared in its accustomed place on
the table. Even when the coals in the hearth grew cold and the
fire died away for the night, the room retained an atmosphere
of reassuring intimacy, the curtains and table rugs soaking up
any stray echoes that would make the space seem chilly and
barren.

I have always had a taste for things of the past—old music, outdated spellings, the history of the Middle Ages, and
not least our own traditional home life—and nowhere did the
past come more alive for me than in the town center of Gouda.

Surrounded and crisscrossed by a network of canals and cobbled quays, Gouda retained much of its character of the sixteenth and seventeenth centuries. The town flourished as a center of commerce even before it was given its charter in 1272. A thriving trade in beer, cheese, and bricks enabled the townspeople to erect a grand cathedral, the present-day Sint Janskerk, as early as the fourteenth century, and to build the striking town hall a century later. At the center of it all lay the market square. On Thursdays and Saturdays, crowds thronged and shuffled among the stalls as vendors bellowed their wares. I drank in the smells of *oliebolen,* delicious greasy balls fried in oil and then smothered in sugar; of fish frying; and of fragrant fruits. The scene would have been incomplete without the carillon of the Sint Jan, the notes of the bells trembling and bending as they dispersed overhead above the din of voices and traffic below. Here was manifested a tangible sense of community, one that even the roar of motorized bicycles and the filling in of many of the canals have failed to dislodge.

Home meant happiness, warmth, and the familiar smell of applesauce cooking or meat frying. Most children would take such serenity for granted, but for me there was an intensity of appreciation. The comforts and continuity of life at home came in small doses, to be savored like pieces of fine Dutch chocolate before they melt away. The years in the hospital were all too soon replaced by years at residential schools for the blind, during which home visits were infrequent and brief, especially by the time standards of a child.

Chapter 3

COD-LIVER OIL

My parents invariably approached important decisions with an enviable level-headedness and an enlightened attitude founded on sound philosophical underpinnings. They instinctively knew that I would flounder in school if I stayed at home until the normal age of six. No, I would attend school at the earliest opportunity.

At this time, there was no compulsory education for blind children in the Netherlands. Those who desired schooling had only specialized residential institutions available to them. A few weeks before my fourth birthday, just months after the final operation on my eyes, I was sent to the Prins Alexanderstichting, a small but very pleasant school located in pine woods in Huis ter Heide, some fifty kilometers from home. During the week, I lived at the school, but on the weekends I was allowed to come home to Gouda.

Huis ter Heide's wooded setting turned out to be ideal for exploration. My first teacher, Juffrouw Mooy, was evidently persuaded that blind children must become acquainted with their natural surroundings. Often she led us across the high trestles of a little-used railway track into a small quiet meadow. Broom grew there, its fuzzy pods plastered along the thin leafless branches; and so did heather, a dry little plant that in late summer would bring forth odorless purple flowers. Never mind if on occasion the prickles of a rosebush got in the way. They were a part of life, to be experienced and understood rather than assiduously avoided lest we should injure ourselves.

Unfortunately, it was decided that the Prins Alexander-stichting would henceforth be reserved for children with residual vision. Those like me who were totally blind would have to transfer to other schools. While Huis ter Heide was under renovation and before a permanent place could be found for me, I briefly attended residential institutions at Den Dolder and Schoorl. Then, at age five, I began school at the Institute for the Blind at Huizen, which we referred to as Bussum because it lay nearer to the center of that suburb than to Huizen itself.

Residential life at Bussum for the younger children was centered in a small two-story building, the school at one end and a row of five attached little houses at the other. All the houses bore the names of birds. I lived in Tjiftjaf (chiffchaff), or House 2, between Pimpelmees and Kwikstaart. Even today, the metal nameplate with the Braille letters too close together still adorns the brick wall to the left of the front entrance, although its lower right portion is now broken away. Twelve children and two housemothers occupied each house then. We slept upstairs in two unheated dormitory rooms, six long oak beds in each, to be made neatly as soon as we rose at seven. The ground floor consisted of a spacious kitchen, used mainly for reheating the meals that arrived from the institute's central cooking facilities, and an L-shaped combination dining and play room, from which the back door led to sandboxes and

a climbing rack made of pine logs. The main part of the
L-shaped room was dominated by a long table lined on either
side with high-backed rattan chairs. My assigned place close to
one end of the table put me next to the ever watchful housemothers, who must have considered me difficult enough to
warrant close supervision.

The short leg of the L comprised an alcove where communal toys were kept and constantly abused. The children hailed
from widely differing backgrounds; some came from penniless
families that rarely if ever looked after them, whereas others
came from homes like mine where life was stable and support-
ive. To prevent even more rivalry than already existed, the institute effectively banned private ownership of toys or anything else. I resorted to keeping what few valuables I owned in
the pockets of my short trousers, which periodically burst as
the pebbles, nails, and pine cones overwhelmed them.

I was as guilty as anyone of treating our communal possessions disrespectfully. All too well I remember an incident involving a book about an old man named Willem who was kind
to children. The book had been laboriously hand-copied into
Braille by a lady who, for unknown reasons, set off every sylla-
ble of every word with hyphens. I found this so silly that I proceeded to tear the book apart, ripping out and shredding a
page or two each day for a week or more. This act of vandal-
ism was committed without misgivings, yet it brought neither
satisfaction nor pleasure. The whole experience at Bussum has
left me deeply skeptical about rules favoring exclusively communal possessions. In the absence of close relatives and loved
ones, I fear, a diminished sense of responsibility and personal
pride allows us all to visit abuse and destruction on the very
things we are meant to value and share.

The school followed the custom of the time to serve the
day's hot meal at the noon hour. Mondays during the winter

months usually meant *zuurkool,* or sauerkraut mixed with potatoes, added to which was the institute's unfortunate specialty of cooked bacon. Meals on other days were equally Dutch; *hutspot* (a tasty mix of mashed carrots, onions, and potatoes), onion-rich stews, *erwtensoep* or *snert* (rich pea soup with sausages), and fried fish. As at home, dessert was usually a thick porridgelike *bloempap* (essentially flour, sugar, and milk heated up), *karnemelkpap* (a lumpy concoction made with buttermilk), or *sagopap* (a similar mixture made with sago). Unappealing as some of these desserts might seem to people today, I liked them—save, that is, for the buttermilk porridge.

Bussum cultivated an atmosphere in which washing the dishes was considered a plum chore. Those like me who habitually ate our meals too slowly were assigned the much more tedious task of drying the twelve sets of plates, spoons, knives, forks, and drinking glasses. I went to great lengths to avoid this kitchen routine. When I began taking piano lessons, I persuaded the teacher to schedule practice sessions on one of the school's instruments from eight to eight-thirty in the morning, just when the breakfast dishes were ready for drying. The ruse worked well for a few days, but like the piano lessons themselves, success was only temporary.

During the months from September through April, the midday meal was interrupted by the odious ritual of swallowing the mandatory spoonful of cod-liver oil. The feeble scarce sunshine of the winter months was insufficient to enable the skin to synthesize vitamin D, a nutrient found in high concentrations in this vile liquid. The housemothers had perfected a technique for ensuring that we would swallow every drop offered. Each victim's spoon was dipped in turn into the ghastly bottle. One might be tempted to leave a bit in the bottom of the spoon, but this merely prolonged the agony, for that same spoon was the implement for dessert. On one occasion, I resolved to try my luck with a fork, but the twenty minutes needed to consume the applesauce that day landed me three meals' worth of dish drying.

As someone for whom social contact was always difficult, I found Bussum a deeply unhappy place. During morning recess, I would steal away to a corner of the schoolyard, preoccupying myself with the gathering and sorting of polished pebbles. So absorbed was I that Wobbe Bruinsma, the gruff headmaster, would roust me out of my hiding place with yet another stern warning to be more attentive to his hand-claps announcing the end of recess. The gadgets and games that entertained the other boys my age held no interest for me. On the days when a pernicious diesel smog settled over the grounds, I was gripped with a sense of entrapment in a world of loneliness and fear from which permanent escape was impossible.

Perhaps most disturbing were the constant violence and strict discipline. Bands of older Frisian toughs would roam the school grounds looking to terrorize us younger boys. I hated their taunts, but I was equally troubled by the punishments the school authorities meted out to these bullies. I witnessed boys being kicked, and the whip was used by authorities on those who had committed offenses as minor as returning to school late from weekends at home. The housemothers, too, were callous. Unable to comprehend Frisian, they strictly forbade children to speak that language. One day a few weeks after she had arrived at Bussum from a small farming community in faraway Friesland, a shy four-year-old girl named Aaltje was sent upstairs in tears for speaking Frisian. It occurred to me that she must have felt even lonelier and farther from home than I did.

It is perhaps not surprising that I found much consolation in religion during these years. At home, Arie and I had been

brought up in a liberal Protestant tradition, and church attendance was sporadic. My father belonged to the sect of the
Remonstranten, whereas my mother had come from the
Doopsgezinden, or Mennonite, tradition. Religion never fig-
ured prominently in family life except at Christmas, when the
story of the birth of Jesus was solemnly read while candles lit
up the Christmas tree. At school, however, I was drawn to the
message of hope and salvation broadcast on IKOR (now
IKON), the interfaith Christian radio service, that aired a few
days per week. I wanted to believe the doctrine that complete
faith in God and his messenger Jesus would give meaning to
a lonely life away from home and family. I yearned for assur-
ance that security and goodness could prevail. The Bible stories of miracles and of God's direct interventions appealed to
me not because I believed them, but because I hoped they
might be true.

Despite this very private infatuation, I cared little for the
compulsory sessions of Sunday school at the nearby Dutch
Reformed church in Bussum. In fact, it was an incident at the
Sunday school that dramatically and completely broke the
spell of religion for me. Each week, all the children brought a
donation of a penny or two. One Sunday, the teacher introduced the ritual of collecting the coins by saying, "Now, let us
give back to God a little of what He has so generously given to
us." As I handed over my penny, the thought went through my
head that God couldn't possibly take or even use the money,
and that instead it would end up in the hands of a church official who would then do with it, well, what? When I posed the
question, the teacher mumbled something about the money's
being a token of our gratitude, that our offering would be
spent wisely for a worthy cause. This explanation seemed
plausible enough, yet a trust of truth had been shattered. If
people could mislead me into believing that I was donating
something directly to God, could they not also fabricate the

other stories about Him? How could I separate fact from fiction? What was I to believe, and on what basis could such belief rest? Perhaps God Himself wasn't real.

This might strike many as an instance of acute overreaction by a naive boy to an utterly inconsequential incident, perhaps embroidered in hindsight. Yet I recall the event and my thoughts about it vividly. Moreover, it came at a time when I had been grappling with the whole notion of fiction in story-telling. Whenever I read anything, I would ask if the story was true or make-believe. Often, I would get the unsatisfactory answer that it was a little of both. What part is true and what part is made up, I wanted to know. The answer was important to me, for I wanted to read only "true" stories, books of fact rather than fiction. If belief in God was based in part on a fictional account, elaborated by people for their own ends, I wanted no part of it.

If I felt lonely and lost at Bussum, I was nonetheless one of the lucky ones who could count on a loving family. Whereas I could go home every other weekend, Mintie van Winkel and Renie van Dijk and many others seemed to have been permanently deposited at the institute, having been all but abandoned by what relatives they might have had. For them, there was no break in the routine, no secure home life to idealize, and certainly nobody like the woman I knew as Tante Greetje.

On those Sundays I spent at Bussum, I would be rescued in the afternoon by Greetje Thedinga, an acquaintance of my mother's from Groningen, who lived a few kilometers away in Naarden. On her bicycle, we would ride into the wooded countryside, sometimes even as far as the shore of the IJsselmeer, and then go to her house to enjoy a delicious supper of fried eggs. While she cooked, I glanced at the intriguing items that lay scattered about. There might be grapefruits from southern France, where Tante Greetje often vacationed, or a new batch of shells from her favorite Wadden island of

Terschelling. These shells always seemed fresher, less beach-worn, than the ones I could find at Scheveningen. Beautifully sculptured bowls and spoons of silver reminded me of home; but it was Tante Greetje's strong Groninger accent that most evoked thoughts of my mother. The *t*'s and *d*'s are formed far forward on the palate with the tip of the tongue, and the *-en* endings of verb infinitives were pronounced with strong emphasis on the *n* while the vowel was swallowed or ignored. Tante Greetje had no children of her own, but she knew perfectly how to cheer up and interest a homesick boy.

As much as I disliked daily life in Tjiftjaf, school was a pleasure. The teachers provided a rigorous education emphasizing Braille, hands-on experience, and independence. The school was not strictly divided into grades. Instead, children were allowed to proceed at their own pace. There were neither examinations nor homework assignments, but our schoolwork was thoroughly checked, and errors were not tolerated. With only a dozen or so pupils in each class, teachers had time to offer individual instruction and to keep tabs on our day-to-day progress.

I learned Braille by means of pegs set in large wooden boards. The system of Braille is brilliantly simple and practical. Its fundamental unit is a six-dot cell, two vertical columns of three, set side by side. Each cell fits neatly under the tip of the forefinger, and when the forefingers of both hands are used, reading is fast and even. Because a given letter or numeral comprises anywhere from one to six dots, sixty-three different configurations of dots can be formed in each cell, more than enough for the twenty-six letters of the alphabet and for the punctuation marks and other commonly encountered symbols. The Braille alphabet has a strikingly logical structure. The first ten letters, *a* through *j*, use only the upper four dots of the Braille cell. Letters *k* through *t* repeat the same sequence as *a* through *j*, with the modification that the lower left-hand dot of the cell is added. The sequence then starts yet again with *u* and *v*, this time with both bottom dots added to *a* and *b*. Were it not for the historical quirk that Braille was invented in France, the next letter *w* would have looked like a Braille *c* with two bottom dots added. There is no *w* in French however, and so *w* is formed like a *j* with the addition of only one extra dot, the one on the bottom right. The letters *x*, *y*, and *z* return to the set pattern, being formed like the letters *c*, *d*, and *e* with both bottom dots added.

This clever and versatile system is easy to master, and many blind people achieve extremely rapid reading speeds, especially because most languages, including Dutch and English, are written with a system of Braille contractions. A single cell can encompass a frequently occurring combination of letters, such as *ea* in English or *ie* in Dutch. Even whole words such as the English *the, like, rather, very, right,* and *these,* and the Dutch *op, van, heeft,* and *geheel,* are compressed into one or two cells.

We learned to write Braille with a slate and stylus. A sheet of paper is secured between two metal plates, hinged at one end. The lower plate is provided with four or more rows of cells, each cell consisting of a rectangle of six tiny depressions into which the point of the stylus fits. The upper plate is a grid or template of hollow cells, matching in position the cells on the lower plate. With the point of the stylus, dots are punched through the paper at the appropriate places to form letters and words. Because the dots emerge on the back side of the paper, each letter must be written in mirror image, and writing proceeds from right to left. The method may seem cumbersome and confusing to a novice, but it soon becomes second nature. If the paper is thin enough, writing with slate and stylus can be almost as fast as handwriting with a pen or pencil. Thick paper gives the right hand and arm a substantial workout and is a little slower, but it is still very practicable. I continue to rely heavily on the slate and stylus to this day. They fit easily into a pocket or briefcase and are less obtrusive to use in a classroom or at a meeting than a bulky, noisy Braille typewriter or a tape recorder. Nowadays the slate is often made of light, durable plastic. The design, convenience, cost, and versatility of this low-tech invention place it near the top of the list of useful implements in my life.

Naive observers, including too many in education, have come to view Braille as a quaint anachronism, a bulky and slow medium from the past that is sure to be supplanted by computer technology and synthesized speech. Nothing could be more wrongheaded. Braille makes blind people literate. It lets us know how words are spelled; it gives us independence; and it enables us to write and revise, to label items, to comprehend musical and mathematical notation, and to work in private. Even with the best of memories, it is hard to imagine how one could master mathematics or a foreign language without Braille. Books on tape or disk suffice for casual reading, provided the listener can stay awake, but they simply won't do for retrieving information or for careful scrutiny. All these advantages of Braille are undiminished for those who can see print but who are unable to discern it with ease. To deny the blind Braille is to deny us literacy and to rob us permanently of opportunity.

The rest of my family also learned Braille. My mother set to work ambitiously hand-copying books for me. There was a wonderful little book about Artis, the Amsterdam zoo, by its director, Dr. Portielje. I read and reread it, each time savoring the names of such faraway places as Australia and California and letting my imagination run with the author as he portrayed the life of camels in the Gobi Desert, fruit doves on lush tropical islands off New Guinea, and condors in South America. Another book consisted of articles from *Tombola,* one of Arie's magazines, recounting the changes in the fields and woods over the seasons as seen through the eyes of a brother and sister. Even Arie and my father copied books for me. One would be waiting at my place at the dining-room table whenever I returned home for the weekend.

My first teachers at Bussum, Juffrouws Frantien and Reilink, strongly believed in acquainting their charges with the outdoors. A steady parade of plants marched across Juffrouw Reilink's windowsills through the year. One day, she demonstrated the fine layering of a hyacinth bulb; later, we helped to plant several dozen. Such simple diversions might strike an onlooker as trivial, yet they are not. The unique benefit of sight is observation at a distance. If that sense is eliminated, nothing at a distance that does not emit or reflect sound or odor can be experienced. Many blind people go through life knowing next to nothing about their surroundings because no one has bothered to describe, or better still to show, the everyday things and circumstances that sighted people take for granted.

One day, I was told that I could advance to Meneer Bontekoe's class. He was a veritable authority figure, with a deep commanding voice and a stern yet gentle disposition. Boys foolish enough to cross him were marched out of the room and given a good thrashing, but such episodes were rare and involved me only as a cowering witness. Bontekoe was a man of very considerable talent who delighted in experimenting with new techniques. It was he who introduced us to the idea of making raised Braille illustrations with a stylus on a sheet of paper laid over window screening tacked to a board. To acquaint his pupils with everyday commerce, he set up a make-believe store at one end of the room. Once a week, each child counted, weighed, and measured groceries and paid for them by counting out and changing real money. The idea was not only that we learn about weights and measures and money, but also that we would know what ordinary items looked and felt like.

History lessons were vividly brought to life as Meneer Bontekoe elaborated the litany of dates and places with stories about the assassination of William of Orange in 1584, life in the towns of fourteenth-century Holland, battles of the Eighty Years War, and the early Dutch explorations of the East Indies. The man had a gift for transporting us back in time.

As a champion of Braille illustrations, Meneer Bontekoe believed passionately in the importance of detailed Braille maps. With the aid of large relief maps showing the main canals, towns, rivers, and railways, he systematically and exhaustively taught the geography of each Dutch province and elaborated the basics by describing landscapes, waterways, and some of the events that had historically shaped the physical geography of the country.

The library room, just down the hall from Meneer Bontekoe's upstairs classroom, housed a collection of stuffed birds. Bontekoe felt strongly that we should know the shapes and characteristics of ducks, swans, owls, eagles, herons, gulls, and the other birds a sighted person could view at a distance. For him, nothing could substitute for firsthand encounters.

Even arithmetic, my most difficult and tedious subject, was brought to life by Meneer Bontekoe. It was taught on the so-called cube slate. A metallic board with a grid of sunken pits held leaden cubes on whose faces Braille numerals had been embossed. Each cube represented a single numeral. Problems could be worked by arranging cubes in rows and columns. On the many winter mornings when the classroom was dank and chilly, my fingers stiffened as I manipulated the icy little cubes, from which the embossed dots had often been rubbed into illegibility. Meneer Bontekoe lightened the routine by pointing out numerical tricks and shortcuts and by engaging us in mental numerical acrobatics.

Twice a week, my courage would be tried to its limits in Meneer Kooyman's gymnastics class. Kooyman quite rightly believed in teaching independence by challenging his pupils to execute difficult acrobatics and to persevere in the face of the threatening situations. He built elaborate structures of rings, horses, ropes, and boards for climbing and balancing high above the mats on the floor. I was mostly terrified by these exercises. Yet, in retrospect, I have come to appreciate Kooyman's efforts, for they prepared me well for the real acrobatics I was to perform on many occasions on cliffs and shorelines around the world.

Life at Bussum was not without its happy moments. I remember well the visit of Queen Juliana and Denmark's Queen Ingrid. The radio broadcast our carefully rehearsed Danish anthem as the queens visited Tjiftjaf. St. Nicolas Day also received ample notice. As the day of gift-giving approached, there would be a celebration in the Koepel. Sinterklass and his helper Zwarte Piet would arrive and throw quantities of the traditional *pepernoten* and other candy for us to scoop up. Palm Sunday, too, was a happy occasion. We would be given so-called palm-sticks, boughs of pines hung with oranges and candies, and even the sullen Meneer Bruinsma would be mellow and full of good cheer.

I look back on these boarding-school years with decidedly mixed feelings. On the one hand, few schools would be as rigorous and as well equipped as were the residential schools for the blind in the Netherlands to teach Braille, to instill self-confidence, and to open doors to the larger world. Without this fundamental grounding, a blind person would lack the tools for fair competition with sighted peers. Yet the school had failings common to many institutions for the blind. Enlightened as it was in teaching the standard curriculum, it made no effort to teach children the use of the cane or any other means of independent travel. Most of us were restricted to the small part of the institute grounds whose layout we knew well, so that it was all too easy to think of oneself as a dependent prisoner kept on a short leash at all times. More important still, the superb education was administered in a totally segregated setting, in a world of the blind totally divorced from the lives of the sighted. Ultimately, the blind must make their way in the sighted world, a society into which they must become fully integrated. The prolonged absences from home were the worst aspect for me. Perhaps I was overly sensitive to the alienation and solitude that this regimented and frequently cruel environment elicited. The discipline, isolation, and self-absorption may have forced independence of mind at an early age, but the price of such uncertain benefit is in the end too high. Adversity would, I think, eventually have claimed me as a victim. Could I have avoided the destiny of weaving baskets or fashioning brooms for a living in the institute's sheltered workshop had I stayed?

The ideal situation, it seems to me, is for a blind child to attend a local school. At first, full attention should be devoted to learning the essential skills—Braille, independent travel, getting to know one's physical and social surroundings—in a class wholly dedicated to the purpose. Gradually, blind children should be eased into class with their sighted peers, with time being set aside to hone and expand the fundamental skills of blindness. Whatever the right solution is for any particular child, the goal of providing the necessary techniques as well as the self-confidence and the social adaptations to live and compete successfully in sighted society must remain clearly fixed at center stage. We cannot hope to achieve this goal either by tossing a blind child unprepared into a class of sighted pupils or by pounding in the fundamental skills in the absence of the social environment in which the person must ultimately thrive.

Chapter 4

THE NEW WORLD

My father had long dreamed of going to America. There, far away from the constant criticism of his father, he might join his uncle Hans Overeinder, who had immigrated to Ohio in the 1920s to start a horticultural nursery. War and Hans's untimely accidental death forced a postponement, but concerns about my education reinvigorated my father's efforts to seek his fortune in the New World. The greatest obstacle lay in finding a sponsor, an individual willing to guarantee employment and financial security for a would-be immigrant.

For a time, my father considered Ohio and Colorado, where distant family relations would act as willing sponsors, but both states educated blind children in residential schools. Neither my parents nor I were keen to prolong the disruptions of family life entailed by schooling away from home. In the end, my father settled on New Jersey, a state in which the newly appointed director of the Commission for the Blind, Josephine Taylor, espoused the enlightened viewpoint that blind children should, to the greatest extent possible, be educated alongside sighted peers in local schools. My father enlisted the help of Herman Schotman, a retired official of the Dutch administration in the East Indies living in New York, to seek a sponsor and to ascertain whether the commission might help with my education. In due course, Schotman found Mr. Walter Stryker, who owned a tract of land known as the Rocker Farm just outside Andover in the northwestern corner of the state. Schotman's letters promised a fully furnished house on the property, together with an opportunity to build a nursery business with Mr. Stryker. Mr. Schotman had seen the house, inspected the fifty-acre farm, negotiated the terms of sponsorship, and enlisted the cooperation of the commission, all apparently in good faith.

With today's cheap communications and jet travel, it is easy to forget the magnitude of the risks and the depth of the unknowns inherent in the transplantation of a well-established family culturally at ease with the traditions of Europe. The realities of a situation and of the personalities involved can scarcely be gauged on paper. Had my father spoken to his sponsor or seen for himself the improbability of the promises made, he would surely have recognized the bait for the dead end it was and reconsidered the bold and irrevocable decision to emigrate. Luckily for me, and I think in the long run for my whole family, he did not.

With the arrangements in place, definite plans for our departure were made in August 1955. We would sail aboard the *Maasdam,* one of Holland America Lines' medium-sized passenger ships, in late September from Rotterdam.

Before departure, however, one hurdle remained. In view of the immigration quotas that existed at the time, officials at the American consulate in Rotterdam were reluctant to admit a family with a blind dependent who would require assistance from the rehabilitation establishment. Immigrants were welcome as long as they did not burden society with costs and obligations beyond the ordinary. An interview at the consulate would establish our eligibility. When my turn came to be interviewed alone by one of the officials, I worried that the application for permanent resident status hinged entirely on my answers. Did I recall anything about my vision, the man asked. Oh, of course, I assured him. I remember all the colors as clearly as if I could see them now, and some shapes too. He seemed satisfied. When approval came later that afternoon, I exulted in the knowledge that I had seen the last of Bussum.

I celebrated my ninth birthday somewhere in the North Atlantic. Nine unforgettable days of rich meals, great ocean swells, my mother's constant bouts of seasickness, and the tantalizing smell of the sea ended as the *Maasdam* glided up the Hudson River on a warm sunny late September afternoon in 1955 and docked at the dilapidated terminal in Hoboken. Walter Stryker's wife, Marion, awaited us and drove the sixty miles, an unimaginably long distance to someone growing up in a small country, to Andover.

It would be difficult to overestimate the novelty of our situation. Language, work, school, food, weather, and the hills and forests of rural New Jersey were all beyond our experiences, and they all had to be absorbed immediately. Except at home, where the family continued to speak Dutch, Arie and I were plunged into a world of English. In Bussum, I had a few rushed lessons, just enough to say, "Ze gramofoon is on ze tebel." My parents had transcribed some introductory English textbooks into Braille. Every little bit helped, but I felt powerless and overwhelmed as I tried to understand others and to make myself understood. Arie's English was only slightly less rudimentary. On his first day in school, he knew that he should return home by school bus, but when eleven buses lined up to transport pupils to all parts of Sussex County, only his acute powers of observation enabled him to find the right one.

Arie could begin school almost immediately, but my entry into the strange sighted world of education was more complicated. First, contact had to be established with the New Jersey State Commission for the Blind in Newark. Within a week of our arrival, we traveled by train to the spacious offices on Raymond Boulevard. Our reception there could not have been warmer. There was an atmosphere of confidence and optimism, the can-do attitude that so often typified America then. On the spot, I was issued a pamphlet outlining the English contracted version of Braille, as well as a cube slate for arithmetic, not the metallic one with leaden cubes familiar from Bussum, but one entirely made of plastic. The numerals on the cubes' faces were so crisp that they irritated the fingertips after protracted use. With an efficiency and purposefulness that would characterize most of my dealings with the commission, arrangements were made for Althea Nichols, an itinerant teacher, to come several times a week to prepare me for Newton Elementary School.

Miss Nichols, a warm and patient woman with a soft confident voice and an unflappably calm disposition, embarked at once on an ambitious program. Not only was my English to be greatly improved, but I had to learn Grade II Braille, the contracted version of English Braille that had become standardized in 1939. All Braille letters are the same in Western languages, but the contractions are unique to each. The symbol for English *th* is an *ô* in French; English *sh* becomes *ie* in Dutch, and so on. Some of the books that my parents had copied into uncontracted English Braille were now retranscribed by volunteers into Grade II, so that I could see for myself how the many rules were applied to real words and sentences on the page.

A second innovation was the Perkins Brailler, a sturdy typewriter invented a decade earlier and still indispensable today. Symbols are formed by pressing combinations of six keys set in a single row on the keyboard. Writing Braille now became virtually effortless if a good deal noisier. Still, Miss Nichols took pains to keep me in practice on the slate and stylus. Just as pencils and pens have not disappeared from the arsenal of writing implements, so the slate and stylus remain indispensable to the Braille user as a simple and reliable means of communication.

Effective participation in a class of sighted children and a sighted teacher mean that I had to learn the keyboard of the standard ink typewriter. Miss Nichols therefore began intensively to train me to type. By the end of October, despite an attack of chicken pox that delayed my entering school, she had managed to pound the basics into my head and fingers.

Meanwhile, the commission approached the principal of Newton Elementary School. Could he persuade one of his teachers to take on a blind boy who knew practically no English? His first target was Ruth Saplow. After earning her teaching credentials nearly a quarter century earlier, Mrs. Saplow had just taken her first full-time position as a third-grade teacher in Newton. Fellow teachers urged her to steer clear of this minefield so early in her career, but she elected to take the challenge head-on. Two years later, she would write an article about her experiences, using Tareeg as my fictitious name.

Mrs. Saplow's classroom provided a vivid contrast with the one I had left behind in Bussum. The quiet, respectful atmosphere in Meneer Bontekoe's room was replaced by a permissive chaos, a frenzy of talking, walking, giggling, and outright disobedience. The academic pace was painfully slow. Immense blocks of time were devoted to art projects, singing, and spelling. There was no history, no geography, and precious little science. I took advantage of the situation by reading ahead in the Braille textbooks that, with the commission's continuing support, had been transcribed by dedicated volunteers.

Whatever the shortcomings of third grade in America might have been, Mrs. Saplow saw to it that I became a fully responsible member of her class. Her sunny, extroverted personality created a forgiving atmosphere in which integration was natural, even inescapable. My classmates never uttered rude remarks about blindness, and the enterprising Mrs. Saplow never met a project or an activity in which I could take no part. When the time came for the class play to be performed in front of the whole school, including my mother, I spoke my lines along with the other would-be actors and actresses, and I played the recorder I had brought from Bussum. *Full inclusion* to Mrs. Saplow was not merely an empty phrase or a distant bureaucratic mandate; it was a state of mind, the manifestation of a deep conviction that the blind should be treated with equality and dignity along with everyone else.

One day, Miss Nichols brought a sleek Braille volume, Brailled on both sides of each page, about the Earth. "Read this aloud to me," she said as she handed it to me. Upon opening the book, I found the first word unfamiliar and unpronounceable, even though it happened to contain one of the new English contractions. "Trilobites," Miss Nichols prompted helpfully, "Continue." The shaky start notwithstanding, the contents of this little book made learning English and contracted Braille almost an afterthought. The text began with a word painting of an ancient sea, populated by improbable creatures whose remains we knew only as fossils. It then proceeded to describe violent volcanic processes that thrust up great mountains and left behind granites and pretty minerals. Other books in the series explored the stars and planets, how sound is made and transmitted, and how water affects the Earth's surface. I consumed them all.

The countryside of northwestern New Jersey provided an invaluable education on its own. I found myself in an endlessly exotic world, in which all the trees, birds, and insects seemed more extravagant than their counterparts in the tidy polders and pine woods I had left behind. My first taste of this exuberance came even as we made our way from the ship terminal to the Rocker Farm. A remarkable chorus of snow crickets ensued, the low vigorous chirps throbbing in synchrony over the wooded land. In the morning we awoke to the plaintive screams of audacious blue jays, their voices so much louder and more insistent than those of the more sedate jays and magpies of Europe.

Large parts of the farm were clothed in second-growth forest. In the pine woods at Bussum, one could walk at will among the trees, but here a tangle of woody vines, many grappling other plants with vicious hooks, impeded access. Snakes and even bears might lurk in these wilds, and poison ivy lay in wait to cover the skin with itchy rashes. Even the more benevolent creations reinforced my sense of mysterious foreignness. Never before had I encountered anything as sweet yet flavorless as the pulpy and improbably soft-textured persimmons that lay scattered on the ground. For a complete contrast, the black walnuts were so hard that nothing could crack them. Their husks emitted a spicy aroma that seems wholly at odds with the edible meat within. Mushrooms, too, were outrageous. Dinner-plate shelf fungi, their woody caps gnarled by thick welts parallel to the edge like exaggerated growth lines on a clam shell, jutted out from tree stumps. Here was extravagant wilderness, the novelty and variety of the creatures made all the more appealing by a pervasive sense of danger.

I had never cared for cold weather, but the discomforts of a damp chilly winter's day in the Netherlands could hardly compare with the onslaughts of the New Jersey winter. Dry, sunny, bitter-cold days and waist-deep snows wore on for months as all the creatures of the forest retreated and withered. The tired coal furnace in our little house sent up a smoky draft of hot dusty air through the metal grates in the floors. Winds howled out of the northwest, and nothing stirred in their path.

Then came the incomparable spring. Eastern North America is blessed with a cast of virtuoso songbirds. How can one top the wood thrush, whose trills and three-note minor-key melodies echo like miniature flute sonatas in the sky? A song sparrow, perched high in a tree at the edge of a field, belts out its trilled overture followed by a staccato of arresting notes, as if to advertise the defeat of the inanimate cold. The unhurried warble of the robin cuts through a spring rain, its major-key triplet bringing intangible optimism as much as does the perfume of spicebush in bloom. Life burst forth, unshackled and seemingly unafraid. The farm's fruit trees sprouted silky tents in which thousands of caterpillars huddled and squirmed. There were sultry hot days when the moist air hummed with insects and hung heavy with the sweet breath of the forest, and then great peals of thunder rumbled over the hills as furies of hail and rain swept through.

Exciting as this new world was to us, Arie and I were not oblivious to my parents' plight. From the day of our arrival, it was obvious that reality bore no resemblance to the promises Mr. Stryker had made to my father. Instead of ambitious plans for gardens, my father found run-down barns on a largely abandoned farm. The only building in good repair was the Stryker house, lorded over by Walter, his shrill wife Marion, and their rambunctious dog Queen. My father was expected to be the caretaker of the grounds while my mother worked as domestic help in the main house. Mr. Stryker was a cold, aloof man with a thin gravelly voice. When he returned home in the evening from work in Paterson, he would lean out of the window of his luxury car and unceremoniously yell, "Eight-t'irty," signifying that my father should report to the house at that time to be told the next day's chores. With a combined income of $50 per week and no independent means of transportation, there was little prospect of improvement, and my parents felt trapped in a rude exploitative world.

To make matters worse, the crate of household goods that had been shipped from Gouda arrived in shambles. Nearly all the dishes were shattered, and several cherished antique chests from Groningen had sustained cracks. Uprooted from their familiar surroundings and deprived of much of their material heritage, my parents had been cast ashore in a land where sink-or-swim attitudes often drowned out a gentler sense of community.

Yet the situation was far from hopeless. Neighbors like Mrs. Jenkins gave us lamps, tables, and chairs and took my mother grocery shopping once a week in Newton. Members of the Presbyterian church in Andover, a village of not more than four hundred inhabitants, introduced us to Thanksgiving with a basket of fruit and a turkey. Another neighbor taught my father how to drive and helped him buy his first car, an unpredictable 1947 Oldsmobile.

With his dreams of working in horticulture now permanently deferred, my father once again turned to office work. With the help of a trusted Dutch acquaintance in New York, he found work as an inside salesman in a small family-owned vinyl manufacturing firm in Belleville, one of the older suburbs west of New York. In July 1956, we moved to a spacious second-floor apartment above Steve's, a snack bar in Dover, where we would spend the next three years.

Dover, a town of fifteen thousand people at the edge of the New York metropolitan area, had at one time bustled with activity as the center of a thriving iron industry, but all that remained now were piles of rock in which fine chunks of

magnetite could be found. Although our apartment stood only
a few steps from Blackwell Street, the town's busy main thor-
oughfare, interesting countryside lay within easy reach. Arie
and I fished in the Rockaway River for suckers and sunfish and
walked miles in all directions to hear the first meadowlarks of
the season, to pick wild blackberries, and to look longingly at
the menagerie of pet birds and turtles in the downtown five-
and-tens.

These casual explorations rapidly turned into a directed interest in all the objects of nature after Mrs. Colberg introduced
me to shells. One day in May 1958, I decided during lunch recess to start a collection of pressed plants. It all began inno-
cently enough. The grassy areas of the schoolyard were dotted
with the rosettes of large plantains, common weeds whose
leaves had a pleasing pattern of nearly parallel veins, each well
separated from its neighbors and distinctly raised from the
smooth surface of the blade. Most of the leaves were either
frayed or had been truncated by lawn mowers, but eventually I
found a perfect leaf, which I slid between the pages of one of
the large fifth-grade Braille textbooks. Soon these ideally heavy
books were bulging with grasses, clovers, and tree leaves. The
fence that ran in back of the grounds was an especially bounti-
ful collecting site, for here grew a profusion of tall grasses and
vines. "There's poison ivy back there," warned Mrs. Fick.
Right she was, but too late, for I came down with a king-sized
rash.

My father was only too pleased to join in the search for
plants. I began to learn from him about leaf shapes, vena-
tion patterns, modes of branching, and flower characteristics,
not to mention the names and characteristics of dozens
of plant families. Every excursion turned into a quest for
green things, as well as for mushrooms, pinecones, feathers,
unusual pods and seeds, and anything else that could be picked

up and examined. Soon I added bones to my collection. At Thanksgiving, I meticulously cleaned the leg and neck bones of the dinner turkey that my father had received from his boss. My mother would buy whole fish for dinner, so that I might extract the backbone. Arie and I even persuaded Mr. Utz, the wood-shop teacher at school, to give us scraps of the many kinds of wood with which he worked. Their varying grains, weights, and odors manifested a surprising diversity.

My collecting zeal even extended for a time to human artifacts. Opa Vermeij and my father habitually smoked cigars, which came adorned with ornately shaped labels. These I pasted in a book, each one accompanied by its brand name. Opa's weekly letters invariably contained a few new choice specimens, decorated with seals and letters in bold relief. My interest waned after Opa died of lung cancer in 1961 at the age of 76.

The company for which my father worked manufactured vinyl coverings for automobile seats, furniture, walls, purses, and the like. These came in hundreds of patterns with names like Crocodile, Snake, Lizard, Bead, Channel, Cinerette, Octagon, Floral, Palm, and Cane. We bound the samples my father brought home into large heavy books, arranged according to textural categories.

Commuting by train in northern New Jersey had become increasingly costly and unreliable in the late 1950s. In an effort to find a place to live closer to my father's work, my parents scoured the surrounding towns for suitable apartments. Eventually, they settled in Nutley, a sleepy middle-class suburb less than a mile from the Belleville firm.

Our little garden apartment was one of those minimally constructed two-story affairs where a spy could have had a field day. The ceiling creaked and thudded with our upstairs neighbors' every move, and it faithfully transmitted snatches of conversation. The heat of the summer sun blazed in through west-facing windows, untempered by air-conditioning. The front door opened directly into the living room, which led by way of a postage stamp-sized hallway into the two bedrooms and the long narrow bathroom. The back windows looked out onto a large field, at the southern end of which an overgrown vacant lot sheltered an intriguing community of plants, ants, garter snakes, and at least six species of land snail. The front faced the tiny garden, where for the first time in his life my father could lavish his attention on growing flowers, trimming the yews and jasmines, and improving the soil.

Nutley justly prided itself on its half dozen large parks, which in many places supported credible remnants of forest. In the midst of endless suburban sprawl, the parks and the streams that ran through them offered welcome opportunities to monitor the seasonal progression of birds, flowers, mushrooms, and even snails. Arie and I also began to grow all kinds of plants indoors from seeds we bought. Every day we measured the height of the seedlings and noted whether new leaves appeared. We never had much luck with radishes; they produced a healthy crop of fragrant hairy leaves, but they never flowered and they failed to produce edible roots. The sunflowers we planted in small clay pots, however, thrived. Their hairy stems shot up to a height of perhaps thirty or forty centimeters before a dwarf yellow flower appeared at the top. Beans also did well, but unlike the sunflowers they were not dwarfed in any way.

The first opportunity to collect my own shells on a New World shore came shortly after our move to Nutley. Cliffwood Beach on New Jersey's Raritan Bay might not win a contest for the most appealing beach, but its coarse sand yielded plenty of surprises for the beginner. As I trod barefoot to the water's edge, jets of water issuing from the sand revealed the presence of soft-shelled clams below. Ribbed mussels, buried in the sand and firmly fixed to bits of sand and gravel by a web of tough fibers secreted by a gland in the mollusc's foot, revealed their presence by the neatly ribbed back end of the shell, which protruded a centimeter or two above the sand. It was surprisingly difficult to excavate the mussels. Under the water, peculiar large animals with legs and a long, sharply pointed tail scuttled away as I laid my hand on them. Horseshoe crabs, the guidebook said. Later, on the ocean beaches of such places as Belmar and Seagirt, I picked up huge valves of surf clams, many still in pairs, thinner and more delicate than the clams I remembered from the Netherlands. More perplexing were flat ear-shaped objects, with a wide spiral wrapped around a blunt knob near one end. They resembled shells except for the texture, which was horny rather than chalky or stony. These, it turned out, were doors, or opercula, that closed off the openings of moon snail shells. These same moon snails left other traces as well, neat circular holes in the valves of surf clams, indicating that the clams had fallen victim to these predators.

The books readily available on shells did not satisfy my curiosity. I wanted to learn much more, and I hoped to see and acquire many more specimens. Suspecting that the American Museum of Natural History must have a substantial shell collection, I wrote the museum to ask if someone there might be willing to identify a few shells for me. The reply from William Old was enthusiastic. Yes, he would be willing to take a look at my shells, and I could even come for a visit. In the spring of 1961, I traveled with Arie and my mother to New York for my first encounter with professional malacologists, people who made their living studying shells and the animals that build them. Riding up the elevator to the fifth floor of the American Museum of Natural History, where the public was normally barred from entry, I could smell the unmistakable odor of preserved birds, old wooden cabinets, and the accumulated dust of decades. As we waited in a large work room, a small man poked his head through the door. Henry Coomans, the collections manager in New York, and later to become curator of molluscs at the Zoological Museum in Amsterdam, was unaccustomed to hearing Dutch spoken. He at once introduced us to William Old, a shy man who, despite the painfully long pauses between words, seemed pleased to show us the collection and to identify a few shells I had brought along. Neither man took any particular notice of my blindness. In fact, Coomans unabashedly encouraged me to study biology and to subscribe to malacological journals. Fossils might be especially suitable objects of study for me, he counseled during a later visit.

With this encouragement, I subscribed to *The Nautilus* and joined the Netherlands Malacological Society. As my mother patiently read every article in the journals to me, I began to see how a scientific inquiry, however mundane, proceeded. Previously, I had assumed that everything I read other than known fiction was true, but the journals showed plainly enough that scientists disagreed with one another and that many earlier interpretations were giving way to new ones. The science portrayed in these articles bore little resemblance to the dry distillation of facts and laws served up in school textbooks. Scientists sometimes made mistakes; their conclusions were sometimes incomplete; and knowledge consisted of an accumulation of approximations rather than of irrefutable assertions. Instead of shaking my confidence in malacology, these revelations gave me hope that I, too, might someday contribute original research. Science was approachable. Many of the questions about shells did not yet have answers. In fact, a very great deal remained unknown, and progress did not always depend on highly technical expertise. Careful observations and simple experimentation could allow even newcomers like me to make contributions to knowledge.

School for the most part portrayed a very different view of science. If there were interesting questions or principles to be explored, the curriculum safely avoided them. On one occasion, the tenth-grade biology teacher worried that if we failed to comprehend the life cycle of mosses, there would be no hope at all for understanding ferns. The big ideas of biology— evolution and the role of DNA as the genetic material, for example—did not even merit honorable mention. Only Mr. Gutknecht's chemistry course offered the thrill of patterns and principles that made clear why people found science a worthwhile discipline.

The lackluster science courses typified a slow-paced and undemanding school curriculum. Most high-school students took only four courses per year. Even the five that college-bound students shouldered left important subjects untouched. I wanted not only all the mathematics, science, and history, but both French and German as well. Counselors and school officials warned that dangerous precedents would be set if I were allowed to take six courses during each of the last two years of high school. I would buckle under the strain, and other students tempted to follow my example might fare even less well. Eventually, however, the principal relented, and none of the imagined disasters came to pass.

Yet this was the age of *Sputnik,* and opportunities to transcend the standard curriculum abounded outside school. In 1963 I was accepted into the Science Honors Program at Columbia University, sponsored by the National Science Foundation. Each Saturday, I journeyed to New York, usually with my father, to join other high-school students in college courses on anthropology, physiology, and statistics. Here at last were deep discussions about evolution and the origins of human beings. Marvin Harris's lectures on Darwin's theories fell on eager ears. Years earlier, when I was introduced to the concept of evolution in Donald Culross Peattie's *Green Laurels,* the concept already made eminent sense. It seemed to explain so much, to organize natural history in so satisfying a pattern. Natural selection demystified our origins, just as probability theory provided the foundation for understanding how the accumulation of rare events could produce complexity and precision over a time interval as long as that during which life has existed on Earth.

The Commission for the Blind not only continued to provide all my school books in Braille, but also paid for Marjorie Young to read to me over a five-year period. While other students frittered away their time in study halls, Mrs. Young read me novels, science, history, travel, and more. She was a tiny woman, very precise in her speech and dramatic in her presentation as she threw herself into the prose.

Although I could hardly have been accused of falling in love with what is somewhat pretentiously called literature, I was not wholly immune to its charms. Reading poetry aloud appealed to my growing sense of language. The rhythm and flow of words and the images they evoked created a kind of mental sculpture whose outlines and structure could be laid bare and embellished with the voice. Thomas Hardy's "The Darkling Thrush" won me first prize in a statewide poetry reading contest in 1964.

With the academic work and my consuming interest in shells, on which I spent hours at home, I had little time and even less inclination to engage in social life. In Nutley I formed a close friendship with Karl Snyder, a quiet serious boy with whom I played chess and horsed around.

In the late 1950s, Arie and I were as caught up as any adolescents in the rock 'n roll craze, eagerly tracking the ups and downs of the popular songs of the day on the hit parade. Then, suddenly and quite unexpectedly, I lost my taste for it. Beginning in 1959, just after I entered seventh grade in Nutley, I discovered a feeling of intense melancholy when the music was played too loud. The incessant beat and harsh off-key singing came to represent all that was so immensely ugly about modern civilization—the noise of traffic, the air pollution, the mindless bustle in huge impersonal cities rife with poverty and filth. Instead one could listen to Schiller poems set to music by Schubert, or to the sonic architecture of any of the hundreds of compositions by Bach. These tastes ensured the permanence of my self-imposed social isolation.

My parents, too, made few concessions to American life. They maintained formal European manners, spoke Dutch at home, and neither invited or visited casually. Cereal and toast penetrated our diet, and we exchanged gifts at Christmas rather than on the fifth of December, St. Nicolas Day, but other American habits were resisted. The blunt, forward demeanor of people offended my parents, who preferred dignified manners and a tidy organized household over the noisy chaos they saw around them.

Unhappiness and disappointment continued to dog my father in his work. Not one to advertise his virtues or to take many risks, he worked slowly and with almost excruciating precision, attending to every detail. He lived continually in fear of dismissal and, although his supervisors appreciated his talents, could never seem to rise to a position of managerial responsibility or power. He seethed privately as fellow employees and supervisors cut corners and slacked off on the job.

My parents returned to the Netherlands in 1973. Even in his new work in Amsterdam, my father found little reward, but this time the chief source of his unhappiness lay in the chronic leg and back pains that increasingly bedeviled him.

Freemasonry gave my father both solace and social contact. He adored its rituals and rose rapidly through the grades, but most important was the respect he enjoyed from others in the Masonic Lodge. Throughout the more than forty years of his tenure as a Freemason, he was repeatedly elected to various posts as secretary and president. In these offices as in his work, he toiled diligently and conscientiously, trusting colleagues to a fault and extending a helping hand to anyone who asked for it. To what extent he embraced Freemasonry's religious symbolism remains unclear to me, but I never doubted that the spiritual dimension mattered more to him than to most of his brethren.

Unlike me, Arie cared little for the academic life. Rutgers University in New Brunswick had reserved a place for him, but he declined it. Instead, in 1963, at the age of eighteen he went to work back in Gouda, for the same company where my father, and Opa Vermeij before him, had also worked. Within six years, Arie landed at the Nederlandsche Middenstandsbank (now ING), where he steadily climbed the ranks to become a highly successful senior training consultant.

I have long admired Arie's decision to forsake college. Even in the early 1960s, society exerted strong pressure on high-school graduates to spend four more years writing term papers, listening passively to lectures, and cramming for examinations. Today those pressures have intensified still further. Arie saw that on-the-job training in conjunction with evening courses in economics suited his temperament better than an unfocused college curriculum would have. Education unquestionably promotes earning power, sharpens our ability to evaluate what we read and hear, and enables us to appreciate the finer things in life; but college is not the only way to achieve these benefits. Intelligent people who are not bookish should not all be forced into the same educational mold. For them, an apprenticeship system built on a strengthened secondary-school curriculum and coupled with problem-oriented training and easy access to education through the public media may provide far greater personal satisfaction and make better use of their talents.

For me, however, the academic path offered the only plausible alternative. Without giving the matter much thought, I had never wavered from my desire to study biology or some related scientific field. At my young age, concerns about employment simply never crossed my mind. Officials at the commission, however, looked on my plans with a jaundiced eye. They pointed out that few fields were more visually oriented than biology and that in any case the battery of tests they had administered to me over the years indicated that a Ph.D. degree might lie beyond my reach. I should attend a small liberal arts school like Williams College, my counselor advised. I might do well there and perhaps even come to terms with the limitations imposed by blindness.

The commission's stance mattered. If my counselors did not approve of my plans beyond high school, they could deny all financial aid, including the funds I would need to pay readers. My parents had essentially no financial resources of their own. I therefore began to reexamine my goals. What about a career in mathematics? I had thoroughly enjoyed geometry and had learned calculus and matrix algebra on my own. Further thought persuaded me, however, that my abilities were no match for the highly abstract concepts that were the chief focus of work in the field of mathematics.

I recalled the advice that Henry Coomans had offered. Why not work on fossils? I could still do all the biology I wanted—in fact, biology lay at the core of paleontology— but, on paper at least, I could de-emphasize those aspects of biology that relied heavily on microscopes or other visual techniques. In reality, of course, I wanted to familiarize myself with those laboratory-oriented parts of the discipline. Happily, the commission agreed to this plan.

Now I had to decide where to begin my career in biology. From time to time I had come across the name of Princeton's vertebrate paleontologist, Glen Jepsen, in the popular press, and from earlier visits I had liked Princeton's small-town setting. Moreover, it had an excellent academic reputation. An

interview with admissions officials at the university early in the
fall of 1964 amply confirmed my favorable impression. I applied to Harvard, Rutgers, and Williams as well, but from the
beginning Princeton was my school of choice.

Even under the best of circumstances, admission to
Princeton from a public high school in northern New Jersey
was anything but assured. Like other Ivy League universities,
Princeton sought a geographically heterogeneous student population. More applications arrived from northern New Jersey
than from anywhere else, so that only a fraction of the applicants from that part of the country could expect admission.
Worse yet, my scores on the standard Scholastic Aptitude Test
hovered near the low end of the range considered acceptable to
Princeton. Then, of course, there lurked the question of blindness. Would Princeton be willing to commit a place for a blind
student, especially for one interested in as unconventional a
discipline as science?

On the positive side, my high-school grades were very
high, and there were a lot of them thanks to my heavy course
load. When principal David Broffman called me into his office
near the beginning of my last year, I was astonished to learn
that I ranked first in the senior class. I now had every incentive
to work hard at the six courses I would take that year. With
this competitive spirit primed, I attacked my academic courses
with vigor and managed to hold on to the top spot.

In April, Princeton and the others said yes. With a full
scholarship in hand, I was about to make a transition as profound as the one to the New World.

Chapter 5

A PRINCETON DIALECTIC

A thick greasy slice of gristly roast beef lay before me like a corpse, mired on a muddy beach of instant mashed potatoes. The overcooked, chopped, canned remains of what once might have been vegetables lay trapped on a tide of gravy. Could I survive here, I wondered as I contemplated my first meal at Princeton's Commons. Perhaps this was an aberration, I reassured myself; but the next day's ravioli, awash in a putridly pungent sauce of tomatoes and cheese, only reinforced a profound culture shock. Gone was my mother's simple yet tasty Dutch cooking, rich in fresh vegetables and heavy on the nutmeg in the green beans, mashed potatoes, cauliflower, and ground meat. The scrupulously clean little apartment in Nutley faded into romantic memory as I faced the reality of a squalid, immensely dusty and drab dormitory room in spartan North Edwards Hall.

My new social surroundings were even more daunting. Many of my peers, having been pampered in moneyed households and groomed and polished in the finest prep schools, affected an air of relaxed sophistication. How could I hope to hold my own with articulate young men who thought nothing of offering guests drinks in their well-appointed dormitory suites, complete with bar and extensive sound system? How could I afford to entertain girls from equally elite settings, who would expect lavish outlays from their hosts for parties and the theater? The parents I met were bankers, lawyers, corporate presidents, even politicians. "My father only makes fifteen thousand a year," said one friend in an attempt to assure me that not everyone was well-heeled. I could not bring myself to admit that my father's pay amounted to half that sum.

Nothing, however, drove home the humiliation more effectively than bicker, the infamous system by which underclassmen were selected as members of one of the fifteen eating clubs on Prospect Street. These clubs were fraternities in everything but name. Campus life revolved in and around them, and the name of one's club revealed at once where in the social hierarchy one belonged. Cannon Club enjoyed notoriety for crudeness and catered to sports lovers. Tiger Inn was for football players; Ivy, Cap and Gown, and Cloister branded members as well-to-do and socially elite. More than 95 percent of juniors and seniors belonged to the clubs. The remaining misfits joined Wilson, the "club for nonclubbies" in Wilcox Hall.

Bicker took place just before the beginning of the second semester. On the first night, each club sent interviewers to rooms where small groups of prospective sophomores waited anxiously to defend their social status. Our inquisitors wasted no time in establishing our suitability as members, nor was subtlety brought to bear in their questions. Where did you go to high school, the more broad-minded of the interviewers would ask. Which prep school did you attend, another more discriminating representative wanted to know. If the interviewer did what mine did, there was no point in talking further. What sports do you like? The question wasn't "Do you like sports?" What are you studying at Princeton? Science? That raised a red flag, for it indicated no strong interest in politics or business. Only five clubs came back the second evening. By the final night, our room had been eliminated by all but the two least selective clubs. I chose Terrace Club, the one nearest to campus, but eventually resigned to join Wilson.

Princeton did not admit undergraduate women until 1969, one year after my graduation. With almost no pocket money to spend, I was not in a position to invite women down for the weekend, even if I knew any to invite. I was in any case shy and completely unsure of myself socially. I looked upon mixers as preventing meaningful social contact. How could one reasonably expect to carry on a conversation when everyone around one was yelling just to be heard over the ugly music? Club parties were even less conducive to socializing. More loud music, a cacophony of drunken, crude people all pursuing a curiously selfish goal of "a good time." The women who enjoyed such decadent, degrading affairs were not for me.

If I found Princeton socially awkward, it was in every other way exactly what I had hoped for. In common with most universities where teaching and scholarship are taken seriously, Princeton boasted a formidable faculty, whose members generally lectured brilliantly and often held positions of prominence in their fields. Physicist Val Fitch, who taught the first semester of the introductory physics course, would receive the Nobel Prize in 1984 for his work on subatomic particles. The second semester of that course was taught by Marvin Goldberger, who later became science advisor to President Johnson.

But what really made Princeton an incomparably rich intellectual experience were my fellow undergraduates. No matter how brilliant any faculty, its powers are limited. For many students, if not for me, professors remained distant figures of authority, with whom contact was largely confined to settings where judgment in the form of term papers, formal discussion groups, and examinations were never far away. Conversations with my peers, on the other hand, carried no such risks.

Daily I encountered remarkable people, with incisive and creative minds, eager to discuss almost anything. I talked music and literature with my close friend Tom Hanson, reviewed physics and calculus with Len Fellman, learned about New York politics from Larry Jay, and read religion and philosophy with Cary Bair. This kind of intellectual life gave real meaning to the word university.

I had carefully constructed a roster of five rigorous courses for the first semester. Besides the physics course designed for those with a background in calculus, I signed up for genetics, accelerated calculus, linguistics, and logic. Advanced placement tests exempted me from introductory biology, and genetics seemed the next logical step. I had already read the textbook, coincidentally written by the father of one of the men in the course, with my mother the previous summer, and taught myself the basics of calculus. This background, I persuaded myself, was sufficient preparation for rigorous physics and biology. I felt confident enough to take the usual two-semester sequence of calculus in an accelerated one-semester form.

To make the situation even more difficult, I elected to fulfill the freshman requirement for physical education with a program of weight lifting. Several afternoons a week, the stringent exercise left me so exhausted that I was unable to work effectively on academic courses in the evening. On top of this already absurdly overloaded schedule, I attended choral rehearsal twice a week as a new member of the Princeton Glee Club. Two motives prompted me to join. First, I loved singing. My second-tenor voice was in no way outstanding, but singing made the choral works of the great masters come alive as no amount of passive listening could. Walter Nolner, the energetic choral director, possessed an infectious passion for Renaissance and Baroque pieces. Singing to the accompaniment of an organ or symphony orchestra gave me a thrill even if the performances rarely met with critical acclaim. My second reason for joining was the promise of a spring tour to Puerto Rico. How could I pass up an opportunity to make my first acquaintance with a tropical island, with its lush forests and diverse shells?

Something had to give. Having been awarded credit for the courses I had taken at Columbia, I could apply for sophomore standing. Because physical education was required for freshmen only, I could drop the heavy load of the weight-lifting program while persevering with the courses and the Glee Club. In early November, the new standing was granted.

Alas, I could not stave off disaster. The grade report dropped through the mail slot of my dormitory room door revealed a four (the equivalent of a D) in physics, and threes (C's) in everything else. Newell Brown, my soft-spoken and highly sympathetic academic advisor, assured me that many first-year students perform poorly. My consistently high grades in Nutley had lulled me into complacency. Now, at the bottom, I faced a crisis. Perhaps a career in science was not in the cards after all. That evening, Terrace Club threw a party for new members. The temptation to drink away my disappointment overcame good sense. But alcohol did nothing to blunt reality; it only reinforced my sense of helplessness and failure.

Fortunately for me, Princeton enjoyed a well-earned reputation of caring deeply about its undergraduate students. Not only did many famous professors teach introductory courses, but students were carefully and conscientiously advised. It was the kind of place where an enterprising student could get to know even the busiest and most renowned members of the faculty.

Within a few weeks after my arrival at Princeton, Newell Brown convened a meeting with two professors to chart my future course work and to consider which kinds of science would suit me best. I could hardly have been more fortunate in the collective wisdom and concern of these advisors. One was Robert MacArthur, who had recently arrived as a thirty-five-year-old full professor from the University of Pennsylvania. Canadian by birth, he spoke softly and with a curiously reassuring monotone and an arresting intensity. As a leading authority in the application of mathematical thinking to biology, he advocated a firm grounding in mathematics. Alfred G. Fischer seemed to agree. At forty-five, Fischer was an eminent paleontologist and geologist with wide field experience in South America and Europe. His rich, deep, highly inflected voice revealed a dignified, thoughtful man who still retained some of the formal demeanor of his German background. Theoretical ecology and evolutionary biology were enjoying a rebirth and, in the view of my advisors, held considerable promise for me. MacArthur and Fischer endorsed the idea of combining biology and geology, both of which enjoyed an excellent reputation at Princeton.

There was, I explained to my advisors, one potential complication. The Commission for the Blind had to approve my course work in biology if I wanted funds from the state for readers. My Princeton advisors were outraged. With their help, I could design a suitable course of study, but the commission had no right to dictate the terms. At once they drafted a letter to the commission, in which they politely affirmed the University's primary role. My counselors at the commission sensibly acquiesced. Never again did they raise objections to my admittedly risky career plans. Indeed, no acrimony of any kind colored their subsequent unwavering support, and my counselor promptly responded with another check.

The flexibility and open-mindedness that MacArthur and Fischer displayed so unreservedly are traits all too rarely encountered in academia. Too often, a well-meaning but inflexible professor both underestimates the power of alternative techniques and clings to unnecessarily rigid protocols. No, you can't take chemistry because you can't do the lab work. True, the mechanics of laboratory practice are for the most part out of reach for a blind person, though there are exceptions; but why should a limitation on manipulation prevent a person from learning the basic principles from lectures or a good book? Even if a student cannot carry out titrations, why can't he or she do the requisite calculations? In many courses, laboratory work involves the examination and evaluation of specimens. Nothing prevents a blind person from participating fully in such exercises. Excellent models of cells, embryos, and three-dimensional molecular structures are available, and these serve as invaluable tools for complementing and illustrating lectures. The use of such alternatives need not imply a relaxation of standards; it does demand flexibility and an instructor's recognition that rigor can be maintained without strict adherence to rules crafted with only the sighted in mind. At Princeton, I was lucky to encounter a nearly unbroken succession of enlightened instructors, who were only too pleased to let me participate on equal terms with my sighted peers.

I was, I believe, the only blind student at Princeton at that time. The university therefore pursued a policy of benign neglect. I recruited readers by making a brief announcement at the beginning of the semester in each class. At $2.50 per hour, students willingly read me the books that they would have had to read in any case. No administrator interfered with these arrangements, nor did anyone prescribe procedures for taking examinations. It was left to me to handle such matters. Today, such independence has often given way to supervision. Many universities have established administrative units that handle and regulate the affairs of disabled students. Readers are found and paid for by the office. Counselors negotiate with instructors about laboratory and examination procedures and in extreme cases even decide whether and how some requirements should be waived. Responsibility has shifted from the individual to the institution. To me this intrusiveness is antithetical to the goal of self-sufficiency. How can one expect to compete successfully in the world if others take charge of the everyday logistics of life and work? Students need training in the alternative techniques that enable them to compete, and they need advice; but they should also be allowed to fail, to learn from their mistakes, to plan, to take risks, and to feel justifiably proud when they succeed by dint of their own efforts.

In my classes I relied on the slate and stylus for taking notes. Nothing could beat their simplicity and efficiency. The commission had given me a tape recorder, but it sat unused in my room. The thought of listening to a lecture twice or to my own whispered condensed version of it was definitely unappealing. In any case, I would have had to transcribe notes from the recordings. Why not short-circuit this process and write down the key points of the lecture from the start, as others did.

My note-taking did not go unnoticed by course instructors. Something clearly perturbed Malcolm Steinberg during the first lecture of his thrilling course in embryology. He paused for a moment, listening and looking intently for clues to the origin of the muffled static issuing from the first row of seats. The pitter-patter stopped. Steinberg resumed his polished delivery, and so did the annoying pops. Finally, he pinpointed the source. "What is that you are doing?" he inquired in a faintly accusatory tone, as he leaned over in my direction. "Well," I replied, embarrassed by the attention, "I am taking Braille notes with a slate and stylus." It was Steinberg's turn to be embarrassed. He hurriedly apologized and picked up where he had left off. If he used the level of background static as a measure of the value of his lectures, Steinberg should have been well pleased.

The professor in an introductory psychology course was not so lucky. He once made the point to his assembled students that he could gauge the quality of his lectures by listening to my dot-punching. The long silences from my hand during many a presentation must have pained him. Alas, I had little taste for the subject, and he was unable to soften the deeply held prejudices I carried from the days when the commission had subjected me to batteries of tests and interviews with earnest but transparently naive members of his profession.

The late 1960s were exciting times in the fields I had chosen to study, and Princeton was one of the centers of the ferment. Robert MacArthur, the leading theoretical ecologist of his day, combined a love of birds with an ability to express the laws of species coexistence in abstract mathematical terms. The number of bird species, he argued, depended on the three-dimensional complexity of the habitat as well as on its area. Species could live together only if they differed by some threshold amount of overlap in the resources they used; otherwise, one would outcompete the others and cause their demise. It was unimportant and even harmful, MacArthur believed, to know the names of the species in question, because the names and the particulars of natural history that came with them obscured the deeper generalities revealed by deduction from first principles. With great skill and clarity, he laid out this world view in a course on biogeography in the fall of 1966. We read the manuscript of his book on island biogeography, written

with Edward O. Wilson at Harvard, which was to be published a year later by Princeton University Press. Each week, he
posed questions for which no satisfactory answers had yet
been found. Why, for example, were there no permanent populations of bananaquits on Cuba? These little birds are ubiqui-
tous on most of the other West Indian islands, and occasional
individuals had been sighted in Cuba, but a breeding population there was unknown. He did not expect a definitive answer, of course, but he did want us to weigh the merits of possible explanations and to design ways of evaluating them.

A course taught by MacArthur alone would have been a
rare treat, but one shared with Fischer was a uniquely rich experience. MacArthur emphasized the role of present-day factors such as area and competition in controlling geographical
patterns of species numbers, whereas Fischer invoked the indelible stamp of history. Harry Hess and others in Princeton's
geology department were in the forefront of reviving Alfred
Wegener's theory of continental drift. Tracking magnetic
anomalies on either side of the recently discovered Mid-
Atlantic Ridge, they confirmed the hypothesis that great continental blocks are moving inexorably apart. Some professors,
like the paleobotanist and stratigrapher Erling Dorf, clung to
the notion that continents occupied fixed positions. Dorf deri-
sively referred to the supercontinent that was supposed to have
fragmented into the present land masses as Euphantamasia.
Fischer, however, was an early and enthusiastic convert to the
new theory of plate tectonics. He pointed not only to the obvious jigsaw-puzzle fit between the coasts of Africa and South
America but also to the otherwise incomprehensible similari-
ties between rock sequences on the two sides of the Atlantic.
As continents moved and seaways opened and closed, barriers
limiting the geographical ranges of species appeared and disappeared. Even the well-known tendency for the tropics to

support more species than the higher latitudes might have arisen from circumstances of the past. Fischer argued that the equatorial regions were more stable over the long run than were the climatically more fickle temperate and polar zones, with the result that the buildup of species had been less frequently and less profoundly interrupted.

Despite their differences in outlook, MacArthur and Fischer pursued their dialectic with obvious mutual admiration. Their debates flowed from evidence and dispassionate argumentation, not from the competitive wrangling that so often drives scientific controversies.

It was in his role as invertebrate paleontologist that Fischer achieved the greatest distinction. In his course on the subject in the fall of 1966, he not only brought an astonishingly rich cache of observations and facts but also created an atmosphere in which unsolved problems and the means needed to solve them took center stage. An especially memorable episode came during a long digression in a lecture about foraminifers, single-celled organisms with intricate skeletons of calcium carbonate. Their abundant remains have long aided paleontologists and stratigraphers in correlating rock sequences from distant parts of the world. A recurring pattern, Fischer pointed out, was discernible in the history of these creatures. A species would suddenly appear in the rock record, persist essentially unchanged for a long time, and then just as suddenly disappear, to be replaced by another species. Such a pattern was not what many evolutionary biologists might have expected. Species should change gradually and slowly over time. Boundaries between species that were separated in time would be artificial and arbitrary if the record of the evolutionary lineage were continuous. This entrenched expectation, Fischer went on, flew in the face of well-established stratigraphic practice. Correlation based on fossils works because individuals can be unambiguously classified as belonging to a morphologically stable and unchanging species, which achieves broad distribution rapidly and then becomes extinct at more or less the same time everywhere throughout its range.

The idea that evolutionary change is concentrated during brief intervals of the geological life span of a species or lineage would come to be known as the theory of punctuated equilibria. When Niles Eldredge and Stephen J. Gould formulated it in 1972, I failed to anticipate the dramatic impact the theory would have in later years on evolutionary biology and paleontology. Fischer had persuaded me that the pattern of brief change and long-lasting stability in the history of species was widespread, so that I simply did not experience the revolutionary nature of Eldredge and Gould's proposal when it appeared six years later.

Long afterward, in 1990, I met the other unsung architect of punctuated equilibria, H. J. MacGillavry, who might well have influenced Fischer. By a bizarre coincidence, this tall Dutchman had come to live in huize het Oosten in Bilthoven, the same well-appointed home for elderly Freemasons where my parents had also settled in 1981. MacGillavry had toured the United States in 1964 to give seminars on his oil-company work on foraminiferal evolutionary patterns, which he published in English in a little noticed paper in *Bijdragen tot de Dierkunde* in 1968. He fully recognized the implications of the pattern, as Eldredge and Gould later did, but he neither sent out reprints nor followed up the paper with additional work. To this day, MacGillavry blames himself for not having been a better salesman for his findings.

Henry Horn's intellectual approach to ecology and evolution was different still. Having just arrived from the University of Washington, where Robert T. Paine was publishing the first results of his famous experiments on the effects of removing predatory sea stars from mussel-dominated intertidal communities, Horn emphasized the power of experiments. It was his interest in the structure of forests, however, that appealed to me most. Horn combined elegant theory with astute observation to show how trees with different architectures competing with one another succeed each other in predictable ways. Leaf shape, wood anatomy, and overall tree form all affect the way a tree cast shade on potential competitors and thus influence the long-term dynamics of forest growth. Horn showed more clearly than anyone else how ecology could be informed by the study of form and function and by the mundane facts of natural history.

Then there was Egbert Leigh. I was introduced to him by John Bonner, Princeton's celebrated and very dignified developmental biologist. I had gone to Bonner for advice on the junior paper I intended to write on spiral growth in snail shells. Bonner modestly told me that, much as the subject interested him, he knew far less about the mathematics of spirals than did the recently arrived Bert Leigh.

In his office, Leigh began as if he had known me all his life. "What the hell is all this about spirals?" he inquired. Before I could formulate a muddled response, he threw out a challenge. "What," he asked, "is the meaning of the triangular coiling of some ammonites?" Bert was referring to the peculiar fact that, contrary to typical snails and coiled cephalopod shells in which coiling follows a smooth regular spiral as the skeleton grows, some Jurassic ammonites coiled in such a way that each loop made three sharp turns to sweep out a triangle. His scientific style was at once clear. If you want to understand a phenomenon, probe its many manifestations and worry about apparent exceptions to the rules.

Leigh soon introduced me to the work of D'Arcy Wentworth Thompson, the legendary British thinker whose book, *On Growth and Form,* combined mathematics, mechanics, untranslated Greek quotations, brilliantly clear prose, and anti-Darwinian diatribes with a deep knowledge of organic form to produce a general theory of how physical forces influence living systems. Over glasses of rum and with Handel organ concerti or Bach flute sonatas playing in the background, we read and discussed Thompson's work on spirals. In his highly idiosyncratic evolution course during the spring of 1967, Leigh offered a heavy dose of mathematics to attack evolutionary problems of adaptation. "Consider a sponge," he began as three dozen students settled in their seats for the first lecture. "Let D be the diameter of the ostium and d the diameter of the osculum. We wish to derive the optimum velocity of the exhalant current in the sponge." Most of my peers had enough. They were not present the next day when, dissatisfied with his first performance, Bert delivered an improved version of the same lecture. As hardy survivors, the remaining students slogged through Ronald Fisher's sex-ratio theory and several other equally demanding topics.

Beneath the mathematical wizardry and the irreverent, often brusque style resided a perceptive, highly informed man with a biological intuition more typical of a field scientist than of a theoretician. He saw beyond the particulars of individual shells and trees to examine whole communities. How did forests growing under comparable conditions in different parts of the world differ in general structure and in the architecture of their component trees? Could something be learned about shell assemblages from the sizes and shapes of the species comprising them? He thought about complexity and stability in thermodynamic terms, and he planted in me, whether intentionally or not, the idea that such complexity must be confronted if we want to understand how living systems work and evolve. He accommodated his biological world view with

Christian faith, which he and his musically talented wife Lizzie practiced openly and seriously. For him, the facts of evolution posed none of the inconsistency and contradiction with Christian claims that I felt. I think religion provided an essential spiritual dimension rather than an explanation for the workings of the secular world and preserved Leigh's cultural links to the Southern family and society from which he came.

The greatest virtue of Princeton's undergraduate curriculum was its insistence on independent study. In the third year (the second in my case), a student wrote the junior paper, a library-based project on some aspect of his major subject. The senior thesis represented a much more ambitious effort that introduced one to the rigors of original research. The designers of these requirements understood deeply that education must transcend passive spectator status in which lectures and preceptorial discussions are the main tools. Nothing beats first-hand experience for teaching the methods of acquiring and evaluating knowledge.

I chose the great variation in shell shape in the common blue mussel as the topic of my senior thesis. Some mussels were flat and fan-shaped, whereas others were inflated and almost cylindrical in form. The lower shell margin, along which the mussel attached itself by means of a bundle of fibers to hard objects of all kinds, varied in profile from strongly concave to straight. Did these differences in shape confer benefits in survival, or did they arise from variations in the conditions of shell growth? How did shell shape vary geographically over the very large range of the species, which in the Atlantic stretches from the Arctic as far south as the Carolinas and southwestern Europe? I planned to answer these questions by collecting and measuring specimens from a variety of habitats—exposed rocky headlands, jetties, sheltered edges of salt marshes, and the like—along the American Atlantic seaboard and in California. Princeton had available research funds for which undergraduates could compete. I was fortunate to secure enough funds to undertake several weeks of fieldwork with a field assistant.

In the summer of 1967, Michael Archer and I flew to San Francisco for two weeks of intensive shore collecting. Mike had just graduated from Princeton as an already highly accomplished student of vertebrate paleontology, and he would continue on to Australia for graduate study. In the years since, he has become one of Australia's leading paleontologists, having made numerous important discoveries of fossil vertebrates. His dry humor, high-pitched laugh, and taste for country and western music accompanied us as we sampled hundreds of miles of California's coast. We braved poison sumac in coastal chaparral, monstrous waves near Big Sur, blowing sand on Drake's Beach, food poisoning at a posh Carmel eatery that probably did not appreciate our disheveled appearance as we washed in for breakfast, and thieves who stole my only pair of wading sneakers in Avila Beach.

Since 1958, I had kept a journal of observations on everything from the earliest call of the robin to details of flowers and bones. I recorded the growth rates of our potted plants as well as my clumsy attempts to classify mushrooms with sponges. During my years at Princeton, the journal slowly evolved into my private forum for criticisms, thoughts, ideas, and observations. I have found that, if I do not write down something when it occurs to me, it will fade away. Many of the entries embarrass me now, but others generated ideas that sometimes ballooned into full-fledged projects, and still others posed nagging questions that cry out for answers.

It was important to make various measurements on shells to describe patterns of variation. No measuring devices capable of achieving an accuracy of 0.1 mm were available to the blind, but Fischer soon thought of an effective modification to the readily available dial calipers. With this instrument, outer and inner diameters are easily measured. The horizontal scale marks centimeters, whereas a circular scale on a dial is marked in tenths of millimeters. Fischer had the horizontal scale notched at the centimeter marks. Then, with the plastic cover of the dial removed, the edge of the circular scale was notched every 0.2 mm. By gingerly feeling the position of the exposed needle on the dial, I was able to ascertain with considerable accuracy the dimension of whatever I was measuring.

I delivered the first draft of the thesis to Fischer in March. Instead of praise, which I had hoped for and even expected, Fischer offered some gentle but persuasive criticism. I should have made some additional measurements and taken into account the work of Norman D. Newell and others whom I had overlooked, he said. Of course he was right. Appropriately deflated, I read Newell's important papers on ancient mussels, measured more variables, and handed in a much improved version. Fischer had once again shown himself to be the great teacher he was. You can't improve without criticism, and teachers must be willing to offer it. The earlier these lessons are learned, the better.

Everyone with hopes of becoming an ecologist, MacArthur maintained, should experience the tropics. After my first encounter with them during the Glee Club's tour to Puerto Rico in the spring of 1966, I knew that the moist tropics exerted an irresistible pull on me. You could sense nature at work there. The bird songs were more raucous, more belligerent; the palms and ferns of the montane forest at El Yunque luxuriated in a fast-paced world in which decay followed on death as quickly as morning does the tropical night. A training grant from the Ford Foundation allowed MacArthur to teach an intensive field course for Princeton graduate students in Costa Rica during January 1968. Charles H. Peterson, a fellow senior who even then displayed his exceptional talent for carrying out experiments in his senior thesis work on the marine communities of New Jersey pilings, was as anxious as I was to join the group. MacArthur readily agreed to take us.

My senses feasted on the rain forest at La Selva. Howler monkeys, dozens of birds, and crickets roared, sang, chirped, trilled, and croaked in a continuous symphony that knew no finale. Huge trees, their bases outrageously fluted into tall thin buttresses, were festooned with a bewildering variety of lianas, some with straplike stems that made helical welts on the host tree, others with mean hooks for snarling branches and fingers.

At my tactile scale, leaves dominated the forest landscape. Some hung vertically, others spread like umbrellas above a vertical stalk, and still others plastered the smooth trunks of trees. MacArthur had pointed out that the leaves of most tropical trees represent minor variations on a monotonous theme: leathery, oval, smooth-edged, with a long drawn-out tip. An observer on the ground, however, perceives a far greater diversity of shapes. Fine velvets cover the leaves of some shrubs. *Cecropia*'s huge ruffled leaves bear a scabrous texture. Most interesting, the climbing plants of the forest possess leaves that depart strikingly in form from those of their host trees. Instead of being oval, their leaves tend toward a heart shape, with a broad base and a long slender leaf stalk. Much more shade would be cast by a leaf of this shape than by the traditional oblong leaf of comparable area in the canopy. I remained interested in the contrasts between tree and vine leaves for years. In 1976, in collaboration with Thomas J. Givnish, I published a paper on vine leaf shape in the *American Naturalist*.

Save for a half hour's stopover in a mangrove swamp near the Pacific coast port of Puntarenas, the course concentrated on forest and mountain habitats on land. Having anticipated this, Peterson and I had looked into the feasibility of visiting a tropical marine site on the return trip from Costa Rica. Together with Michael Berrill, one of MacArthur's senior graduate students, we traveled to the Caribbean Marine Biological Institute on the Dutch Leeward Island of Curaçao.

I have never fully understood why Curaçao and the other nearby islands and the coastal region of Venezuela are so dry, but their sparse vegetation of cacti and xeric trees accurately mirrors a weather pattern of strong northeast trades and infrequent rains. Lizards, *Cerion* land snails, and reedy cicadas make this desert island their home, as do bananaquits, flamboyantly singing mockingbirds, and the more refined low-voiced oriole. There is exuberance here, but not the feeling of omnipresent life that envelops the body and fills every sense to capacity in Costa Rica.

The shore in front of the Institute consisted of a jumble of coral boulders set in coarse sand. Most of the visitors to the lab head for the reef and scarcely notice that these shore rocks also support a diversity of living corals. Freek Creutzberg, the Dutch director, warned us about the long-spined *Diadema* sea urchins that lurked among the boulders. Their spines, he said, not only were long and very sharp, but they waved almost as if to seek out and sting the hand that casts a shadow over the urchin. Having already become acquainted with sea urchins in California, I shrugged off his warning as an exaggeration. When the shock of the needle jab subsided as my finger throbbed, I understood the problem. Bits of the spine break off and lodge in the skin, kept there by tiny barbs. I proceeded with great prudence, groping among the rocks. Here at last were some of the tropical shells I had read about in books and admired in museum drawers. Large, flat, round *Isognomon* clams hung from mangrove roots, small neatly ribbed mussels tucked in among them. Many small snails, their shells delicately sculptured with beads and ribs, sheltered under the boulders, while other larger species with thicker shells, their surfaces often thickly coated with the limy growths of coralline seaweeds, rested exposed to the sunlight.

Back in Princeton, I divided my time between the thesis and the newly acquired shells. To my astonishment, I discovered that the few *Cerion* snails I had plucked from the limestone outcrops on the island had eaten the paper labels I had written. These snails spend much of the year as inert white ornaments on the landscape, becoming active only when wet. Now, having been disturbed and briefly washed, they had awakened for a quick meal, ready to resume torpor.

I knew in my heart that natural history would play a leading role in my future in biology. Theory was powerful and enticing, but I lacked the mathematical creativity that theoretical advances would demand. Moreover, I felt uneasy about MacArthur's insistence that names and details don't matter. To ignore them was to throw away useful information, to cut off lines of evidence that might profoundly affect the big picture. I might never have arrived at this disagreement with MacArthur had it not been for his encouragement to wonder and to ask. With MacArthur's shocking and untimely death of cancer in 1972, I lost a friend and a hero. But the others thrived, and they have continued to influence my thinking to this day.

Chapter 6

THE OLIVE AND THE HARP

In the fall of 1967, the time had come to plan for life after Princeton. I was intent on pursuing doctoral studies in biology and geology, but I had thought little about which programs would be most suitable or even whether they would take me. Most of all, I wanted to do research. Programs emphasizing course work were unattractive, as were those that imposed a large number of requirements. I was looking for freedom in an intellectually challenging environment.

For a time I toyed with the option of returning to the Netherlands. Like the rest of my family, I found many aspects of American life in the sixties unappealing. Save for oases like Princeton, much of the country seemed to disdain intellectuals and to revel in the banalities of television. Equally unsettling were the racial strife in the cities, the popularity of drugs and the student uprisings on college campuses, and the squandering of lives and money in Vietnam. All the ugliness and decay of the age were captured and amplified by the dissonant music that blared intrusively out of dormitory windows. Naively, I believed that Europe still clung to a higher standard of refinement.

Thoughts of a European education were short-lived. Letters to the geology department at one Dutch university elicited the quick response that the study of paleontology by a blind person was totally out of the question. A professor at Leiden replied to my inquiry by stating that, although he had no objections to my studying with him, his type of research on organic form might not suit my interest. At the time this sounded like a lame excuse, but in retrospect I think he was right. I also considered Cambridge briefly, but Bert Leigh persuaded me that I would likely feel lonely and not well served there.

Having decided to stay in the United States, I turned my attention to several eastern universities that had close associations with museums. Harvard's Museum of Comparative Zoology was unusual in having not just one, but several distinguished curators of molluscs. Many of the malacologists in America had been trained there. Then there was Columbia University, which was close to the American Museum of Natural History. I knew little about the university except that two very fine paleontologists, Norman Newell and Roger Batten, taught there and had close ties with the museum.

My first choice, however, was Yale. A paper by A. Lee McAlester on *Babinka,* a primitive bivalve that appeared to link early segmented animals with the more derived clams, caught my attention in a newly published issue of *Malacologia,* a journal to which I had subscribed since its beginning in 1962. McAlester held an appointment not only in geology at Yale but also at the university's Peabody Museum. Two others from the museum, Copeland MacClintock and Giorgio Panella, were engaged in exciting work on daily growth patterns in living clams. Their joint seminar at Princeton fired my imagination. Quahog clams, their work showed, laid down an average of one growth increment per day. Clams three hundred million years old had roughly 395 such increments. Did this mean that the Paleozoic year had some thirty more days than years do today? Here was a dramatic demonstration of the power of fossils to chronicle unexpected events and circumstances in the distant past.

Philip Gingerich hoped to study vertebrate paleontology and also had his sights set on Yale. The son of an Iowa dentist and a person of singularly independent disposition, he had entered Princeton as a football player with plans to study mathematics, but he soon discovered the allure of geology and henceforth devoted himself with vigor and great talent to paleontology. Together, we arranged for a day of interviews in New Haven. We were convinced that only through face-to-face meetings could Yale judge us adequately and we see how reality stacked up against reputation.

My first appointment reassured me. Lee McAlester, a quiet Texas gentleman, brimmed with enthusiasm over my work with mussels. Steven Stanley, who had recently completed a brilliant thesis on the correspondence between form and habitat in living as well as fossil clams, had studied with McAlester. That kind of success story reflected well on McAlester and on Yale generally.

The main event of the afternoon was to be an interview with Edgar Boell. Comfortably housed in his eleventh-floor office in Kline Biology Tower, this grandfatherly and very gracious director of graduate studies in the biology department radiated an air of quiet, benevolent authority. His skepticism was palpable. How, he began, could I work morphology? How were my grades? Did I have interests besides biology? How could I possibly stay on top of the vast scientific literature in my or anyone else's field if nothing was published in Braille? Patiently, I explained my methods, recounting my experiences at Princeton and my long-standing dedication to the study of shells. Boell listened intently, but I sensed that his doubts wouldn't yield.

Would I like to see the mollusc collection? Boell asked. Indeed, I would. Down the elevator, down the long flight of outside steps, and into the basement of Peabody Museum we went. Percy Morris, the curator, was there to greet us. Like Boell, he must have been in his sixties, and he tempered his welcome with a touch of wariness. If only they don't ask me to identify anything, I thought to myself.

"Here's something. Do you know what it is?" Boell asked as he handed me a specimen. My fingers and mind raced. Widely separated ribs parallel to outer lip; large aperture; low spire; glossy; ribs reflected backward. "It's a *Harpa*," I replied tentatively, "It must be *Harpa major*." Right so far. "How about this one?" inquired Boell, as another fine shell changed hands. Smooth, sleek, channeled suture, narrow opening; could be any olive. "It's an olive. I'm pretty sure it's *Oliva sayana*, the common one from Florida, but they all look alike."

Both men were momentarily speechless. They had planned this little exercise all along to call my bluff. Now that I had passed, the time had come for me to call theirs. Boell had undergone an instant metamorphosis. Beaming with enthusiasm and warmth, he promised me his full support. Yes, he said, he had had his reservations, but I had convinced him.

It takes a certain security of mind, a lack of pretense, to admit error and to change one's mind as radically and as openly as Boell did on that November afternoon. What would have happened if the director had instead tried to save face and clung to his prejudices?

The interview with Willard Hartman, whose cluttered smoky office was just down the hall from Morris's lair, proceeded in an anticlimactic atmosphere. After telling me a bit about the sponges whose study he made his life's work, Hartman shyly inquired about my blindness. Still, optimism got the better of him, and he appeared willing to offer me the chance to prove myself in graduate school.

I had reason to be guardedly optimistic. After the first disastrous semester at Princeton, my grades had improved dramatically, with the result that I would rank in the top five percent of my class. My academic advisors had confidence in my enthusiasm, and I was pleased with the outcome of the interview in New Haven.

Acceptance to Yale in February with a full fellowship made the rejections from Harvard and Columbia almost irrelevant. Both schools informed me that they lacked the special programs and equipment that blind students needed. Harvard in particular considered work on shells by a blind student outside the realm of feasibility. Still, their rejections drove home an important lesson. Preconceptions about the abilities of the blind are widespread, and only the dispassionate and unconfrontational presentation of evidence face to face, together with a reasonable self-confidence, can overcome such prejudices in people with an open mind.

Shortly after the Yale interview, yet another open-minded person who was to influence me profoundly came to Princeton. Robert MacArthur had invited Alan J. Kohn from the University of Washington to present a seminar about his work on cone shells, an extraordinarily diverse group of more than four hundred species belonging to the genus *Conus*. When Kohn entered Yale to become one of G. Evelyn Hutchinson's graduate students, among whom MacArthur was another, he decided to tackle the ecology of these snails. Although their elegantly marked conical shells had been sought avidly by collectors for more than two hundred years, practically nothing was known about their biology. Kohn soon discovered that as many as a dozen species could be found living side by side on reefs and rock benches in Hawaii and other islands in the tropical western Pacific and Indian oceans. How, he asked, do these animals, which differ among themselves mainly in subtleties of color and form, divide up the physical habitat and the available sources of food? How do they find and ingest their prey: worms, snails, and in a few specialized members even small fish? Kohn, whose thesis ultimately grew into a lifelong commitment to *Conus,* was among the first scholars to learn something about these animals other than their names.

In the seminar, he spoke with quiet yet palpable passion about how *Conus* speared their prey with specialized venom-containing teeth released from the ribbonlike tongue or radula. Species that ate fish, Kohn remarked, typically have a wide shell aperture, whereas worm-eaters have a very narrow opening.

The wide aperture presumably enables the snail to swallow its envenomated victim whole. Coexisting cones showed little overlap in the species they ate and also occupied slightly different microhabitats. In California, where only one species of *Conus* occurs, a great diversity of prey species is taken by that one species.

Here was a man who combined a love of shells and tropical shores with the larger aim of discovering broadly applicable ecological principles. Over lunch after the seminar, I bombarded him with questions. Recognizing that my enthusiasm was genuine, Kohn mentioned that he, along with three other malacologists, would be offering a course on molluscs the following summer at the Hawaii Institute of Marine Biology on Coconut Island. Would I be interested? Indeed I would.

Once again, acceptance was problematic. Not only would I be the least experienced and youngest member of the class of twenty, but officials in Hawaii raised doubts about whether I, as a blind person, could benefit from a program in which so much depended on being able to see. Thanks to Kohn's intervention with Philip Helfrich, the institute's amiable and open-minded director, I was finally admitted in April.

Helfrich had assembled an impressive cadre of malacologists. Alan Kohn represented the ecological perspective and also had a lively interest in how molluscs, including their shells, function in nature. The University of Hawaii's Allison Kay not only was an expert on the natural history of Hawaii and other Pacific islands, but she was also in the midst of preparing a complete guide to the Hawaiian molluscan fauna. She had an extraordinary familiarity with the full range of the diversity of molluscs, especially including the hundreds of small species, and also understood something of the biology of cowries, another group of snails favored by collectors but otherwise virtually ignored by biologists. Her enthusiastic sparkle brought this knowledge alive and made it accessible to anyone who cared to tap it. Vera Fretter from the University of Reading was one of the world's leading molluscan anatomists. With a wry sense of humor and frequent impish laughter, she set out to dissect all manner of snails. Martin Wells, a strikingly articulate neurophysiologist from the University of Cambridge, rounded out the faculty. He was a leading student of the behavior and physiology of *Octopus* and would later go on to conduct important studies of locomotion and evolution in this and other cephalopods.

For most of that summer, my home was a small spartan barracks on Coconut Island. This little world, the sound of whose Hawaiian name, Moku O Loe, much better fits my image of a tranquil tropical oasis, is situated in Kaneohe Bay on the cool windward side of Oahu. Its hilly center, alive with introduced doves and mynahs, is surrounded by an artificial portion on which the laboratories and living quarters were built. The damp northeast trade winds kept the fronds of the coconut palms in constant motion and the water ceaselessly splashing against the dock a few paces from the barracks. The air was so moist that the Braille paper I had brought with me took on the texture of a limp rag, which proved surprisingly resistant to the stylus and the embossing head of the Perkins Brailler. Inside the barracks, geckoes had staked out territories on the ceiling, and large noisy cockroaches couldn't decide between dinner in the kitchen or a meal in the bowls of shells drying beneath my bunk. Even toads occasionally wandered in, no doubt eager to feast on the many well-fed insects. This peaceful if unruly regime was often interrupted by the deafening roar of military jets as they swooped overhead or exercised their engines on the runways of the nearby Kaneohe Marine Air Station.

The cockroaches under my bunk had reason to be pleased. Many of the snail shells I was collecting had apertures so small or so constricted by thickenings projecting from the sides of the openings that extraction of the soft parts with a needle proved difficult. Often, the snail's foot had withdrawn to a point beyond the reach of the needle. I gave the matter little thought at the time, but it was clear that these Hawaiian shells were considerably harder to clean than those from Curaçao, California, or New Jersey.

During the second half of the course, each student was required to carry out an independent research project. After several false starts, I settled on the ecology and shell form of a group of river-dwelling snails belonging to the family *Neritidae*, most of whose other members are marine. Several curiously cup-shaped nerite snails were found exclusively in the fast-flowing streams of the Hawaiian Islands, and these differed in interesting ways from their marine relatives, which cling by the hundreds to rocks between tidemarks on warm shores. The fresh-water species had much larger apertures, a much thinner oper-culum or door that closes off the aperture, and the odd habit of laying eggs on one another's shells.

Although the Hawaiian fauna technically belongs to the much larger Indo–West Pacific biogeographical province, its isolation from other parts of the western Pacific and Indian oceans means that many typical species that are widely distrib-uted in this province are absent from Hawaii. Having come to this outpost of the world's richest marine region, I very much wanted to take a brief look at an island whose fauna was more typical of the Indo-Pacific. Besides, such a place would also harbor a number of freshwater nerites that Hawaii lacked. Alan Kohn and Allison Kay concurred and recommended Guam. They wrote to Lucius G. Eldredge, who three years ear-lier had finished his doctorate on the classification of tropical sea squirts and who was now teaching at the College of Guam. Lu's response was guarded. He reminded us that Guam's shores were thick with *Diadema* as well as with other danger-ous animals such as poisonous cone shells and stonefish. How was I prepared to handle myself in such situations? I assured Lu that *Diadema* and I had already met and that nothing unto-ward had happened to me so far in Hawaii. He invited me to come. I could stay with him, as long as I could put up with four young daughters, several cats, and sleeping on the lanai.

From his letters, I might have surmised that Lu and his family were reserved, perhaps even a little protective. Nothing could have been further from the truth. The Eldredge household—Lu, his energetic wife Jo, their four daughters, and two Siamese cats—exemplified to the fullest imaginable extent the optimistic attitude that practically nothing was im-possible. Do you want to see the wave-swept reef margin? Sure, why not. Talofofo Falls is hard to reach on a slippery, steep, muddy path, but there might be some very interesting freshwater snails there, so let's go. Did you say you were leery of hot foods? Surely not; you shall have some chicken kelugen, with the hottest red peppers and the finest lemons we can find, and you shall like it. So I did.

Guam was love at first feel. An exquisitely humid blanket of warm air, laden with the scent of plants and a touch of mold, enveloped me as I emerged from the plane before dawn that August morning. Here at last was an unmistakably tropical island, where even the everyday creatures were extravagant and audacious. No more timid snickers of the geckoes at Coconut Island; the geckoes in Lu's airy one-story house on Dean Circle cackled loudly while high-pitched crickets chirped at a feverish pace.

A glimpse of the undiminished fauna of a western Pacific island is what I had come for, and I was not disappointed. On the first afternoon, Lu drove me a few miles south from his house to a shallow broad embayment known as Togcha Bay, on Guam's windward east coast. A vast bench—an expanse of limestone, shallow pools, and little gullies extending out from shore as a reef flat on which one could walk at low tide—lay teeming with an incredible diversity of life under the broiling sun. At the landward edge of the reef flat, sharp outcrops of limestone were pockmarked with crevices in which we found diminutive spiny periwinkles and small, curiously high-spired nerites that resembled periwinkles in shape. I did not know it then, but the spiny periwinkle would turn out to be an undescribed species endemic to the southern islands of the Marianas chain, of which Guam was the oldest and largest; and the nerite was *Nerita guamensis,* described in the early nineteenth century by Quoy and Gaimard, two perceptive and brilliant French naturalists, and then more or less forgotten. Nearby, where the tide wet the rock a little more often, finely sculptured limpets, with a sharp apex centered above an almost perfectly round base, had excavated deep cavities in the limestone. These, it would turn out, also belonged to a new species, which David R. Lindberg and I named *Patelloida chamorrorum,* in honor of the Chamorro people of Guam, in 1985. The thunderous roar of the surf beating on the edge of the reef drew nearer as we headed out across the bench. Pools left by the receding tide teemed with snails and hermit crabs, which seemed not to mind the heat of the sauna in which they grazed. Lush meadows of short turtlegrass, its leaves peppered with tiny green nerites, passed beneath our feet. I picked up a small conch that was partially buried in the sand. It was readily identifiable as *Strombus gibberulus,* but I was about to discover an attribute that familiarity with the shell alone cannot convey. The snail kicked violently with its agile foot, which was provided with a long, curved, clawlike operculum with saw-toothed edges, or margins. Clearly, not all snails cower in their shells when disturbed.

As we neared the edge of the platform, we came upon a dense bed of *Sargassum,* a coarse brown seaweed whose leaflike fronds bore sharp crinkly margins. The water at our feet was now refreshingly cool, and it rushed to and fro with each incoming wave. Finally, we reached the reef margin. All the rocks were thickly encrusted with corallines, red seaweeds whose calcium carbonate skeletons form a wave-resistant veneer. Astoundingly flat limpets, their edges deeply scalloped, formed tiny excavated territories on the corallines as the result of laboriously grinding the rock with their radula, a tonguelike ribbon of teeth. The topography was no longer the featureless flat bench. It was deeply fractured with innumerable surge channels and crevices, lined with sturdy coral heads. Turban snails, along with large conical top shells, and glossy cowries clung tightly to the rocks in the holes and crevices where aggressive crabs and moray eels also lurked. Three or four species of drupe snails, with a massively thick knobby shell and an impossibly small opening lined with teeth, were somehow able to hold on as wave after wave swept over them. Their apertures were so constricted and the shell so thick that there hardly seemed to be any room for the snail's body inside. Occasional patches of a delicate, green seaweed dotted the rocks. A tiny thin-bodied crab always hid in the seaweed, never itself being exposed to potential enemies. When all the other life-forms were so sturdy and so heavily mineralized, how could these apparently undefended creatures survive here?

As the tide turned, more and more water was coming over the reef crest, and it was time to retreat. It would not be the last time that Lu gave me one of the great biological thrills of my life.

Back in Hawaii, I had time to reflect on the three warm-water faunas to which I had been introduced during the past eight months. Each was, of course, interesting in its own right, but it was the comparison among the three that revealed the most curious patterns. The habitat I had sampled most intensively was what marine biologists call the upper shore, a place where incompatible marine and terrestrial influences come together in an uneasy standoff. Around the world, this spartan environment is occupied by a highly specialized cadre of snails. On tropical shores, these creatures must withstand not only long dry spells when the tide fails to reach them for days on end, but also downpours of rain, unrelenting hot sunshine, and a chronically low food supply. There were at least five species of periwinkle in Guam and eight in Curaçao, but only three in Hawaii. Those living in the highest zones in Guam and Curaçao were either strikingly high-spired and subtly sculptured with a dense cover of fine spiral threads or more compactly built and beset with closely spaced small beads or prickles. Hawaiian periwinkles were plainer. The one species that was beaded bore weak sculpture compared to its counterparts in Guam and the Caribbean. There was an equally interesting pattern among the nerites. Guam's fauna contains at least seven species, ranging in shape from the periwinkle-like *Nerita guamensis* and the spherical, heavily ribbed *Nerita plicata* of the upper shore to the flattened *Nerita albicilla* of the middle shore. The four West Indian nerites show a smaller range of forms; the two high-shore specialists are less spherical and less strongly ribbed than their Guam counterparts, whereas the two midshore species are more finely ribbed and not as distinctly flattened as *Nerita albicilla* in the Pacific. Only two nerites are common in Hawaii; both range widely over the shore, and both have a shell shape intermediate between high-shore and midshore specialists. A third species, *Nerita plicata,* is occasionally encountered in the upper zone, but the Hawaiian specimens are less globose than those from warmer islands such as Guam.

Several aspects of these patterns were intriguing. Snails from the uppermost reaches of the shore seemed to be predictably different in shell form from those found in lower zones. Did these differences reflect predictable patterns in the factors to which these snails had responded over evolutionary time? Perhaps the rigors of drought, heat, and low availability of food created selection in favor of only certain shell characteristics, whereas greater exposure to waves and the more moderate conditions of the lower shore favored a subtly different array of possible shell types. Most interesting, why was the apparent degree of habitat and shape specialization greater in some places than in others, and is this specialization in any way related to the number of coexisting species?

Here, at my fingertips, lay the raw material for a possible Ph.D. dissertation. I could go to Yale with the knowledge that research in comparative biology lay in my future.

Chapter 7

THE HOT HIGH SHORE

If one wanted to work at the intersection of biology and geology as I did, Yale was the place to be in the late 1960s and early 1970s. Willard Hartman, my reflective and soft-spoken advisor, was a world authority on living and fossil sponges and possessed seemingly infinite knowledge of all the other groups of invertebrates as well. G. Evelyn Hutchinson, the erudite elder statesman of ecology, had not only published definitive treatises on limnology, the study of fresh waters and the life in them, but also had a large role in shaping the field of biogeochemistry; after all, he wrote an entire book on vertebrate excretion. He had also achieved an early understanding of the evolution of crustaceans and other arthropods. His breadth of knowledge and understanding was enormous, and he was more than happy to point willing listeners to obscure papers on every imaginable subject. "That's very interesting, but have you thought of . . . " was his refrain in nearly every conversation I had with him. Never did he utter a word of discouragement. He, like Hartman, believed ardently in the independence of graduate students, and he maintained a hands-off attitude that to me was a sign of confidence and trust.

On the more strictly geological side, Donald C. Rhoads was investigating how the burrowing and feeding activities of marine animals affect the muddy bottoms on which these creatures live. A. Lee McAlester, whose work on early bivalves had already attracted my attention in Princeton, was fascinated by large-scale patterns of ecology and diversity in the fossil record. He gave freely of his time and research money to help me get started. It was through his efforts that I acquired a little office in Kline Geological Laboratory, where I could interact on a daily basis with other graduate students who shared interests in paleontology.

However distinguished the faculty might have been, it was the graduate students who gave Yale its unique intellectual vigor. Seminars organized by McAlester and by Charles Remington, the biology department's evolutionary biologist, featured spirited exchanges among the likes of Robert C. Cooke, Alan Covich, Jeremy B. C. Jackson, Jeffrey S. Levinton, David L. Meyer, and Charles W. Thayer, all of whom would become leading scientists in their fields. I probably learned more about fossil vertebrates from my fellow graduate students Philip Gingerich and Peter Dodson than from any formal course.

The layout of the geology building contributed to an atmosphere in which graduate students from different disciplines conversed freely and often. Just a few steps from the wing that housed our little offices was the Dana Club, a room set aside for graduate students, where one could sit around large slate-topped tables to drink cheap coffee, have lunch, read a newspaper, and gossip. Even professors often dropped by for a chat. It is a pity that, in an effort to squeeze all the offices and laboratories possible into the limited space available in most departments, such places for casual gatherings are the first to disappear. I believe they contribute immeasurably to the intellectual health of academic departments.

What I sought most for graduate school besides a lively intellectual environment was the opportunity to do research. Before I could devote my full energies to this enterprise, however, I needed to fill some educational gaps and to pass a two-day written comprehensive examination. I resolved to sweep these requirements out of the way by the end of the first year at Yale.

There are areas of biology where understanding hinges on knowing a huge amount of information. Invertebrate zoology is one of those disciplines, and Willard Hartman's knowledge surpassed everyone else's. No group was too obscure or too recently discovered to escape his notice. Quietly, methodically, and thoroughly, he delivered a torrent of detail in almost deadpan fashion, choosing his words carefully and slowly. His office was beyond clutter. Stacks of books competed for space with jar after jar of sponges, and a pall of cigar smoke hung over it all.

Hutchinson's encyclopedic knowledge came packaged in well-crafted lectures, delivered in a pleasing Cambridge accent in soft, almost murmured tones. There was a polished formality about them that perfectly matched his gentle nature. Hutchinson emphasized the importance of competition in regulating the sizes and densities of populations, yet he himself was as uncompetitive a soul as one could ever hope to meet in academia. Frail and bent over he might have been, but his intellect showed no sign of age other than his frequent references to older and largely forgotten literature, from which he did not exclude the classics or the writings of medieval scholars.

Although I was housed in geology, I had enrolled in biology and therefore fell under its rules. Students were admitted

to candidacy upon the successful completion of the comprehensive examination. I saw no reason to postpone this rite of
passage to the end of the second year, as was customary, and
therefore elected to take it in May 1969. One of the department's secretaries dictated the essay questions. As I typed them
in Braille, my hands shook uncontrollably. Five of ten questions were to be answered in great detail, and all ten elicited
waves of doubt and confusion in my mind. What were the factors regulating populations? Why was the worm shape so frequently evolved among invertebrates and even among many
snakes and lizards? What were the five great advances in the
history of the plant kingdom? Somehow I cobbled together the
answers; somehow I passed.

I was now free to concentrate on the dissertation. Much of
my time during the first year not spent in courses or in preparation for the comprehensive exams was devoted to background
reading about the natural history of rocky seashores. In the narrow belt of habitat affected by the rise and fall of the tides,
plants and animals must cope with exposure to air, sun, and
rain, conditions to which living things on land are well adapted,
as well as with immersion in seawater. At the upper reaches of
this intertidal belt, the marine influence is small, but most of the
inhabitants are marine organisms that are specialized to endure
the hardships of life out of water. Farther down the shore, the
marine influence becomes stronger, and the diversity of marine
life rises sharply. At and below the low tidemark, we find the
undiminished variety of creatures that we associate with the sea.

The Society of Sigma Xi had awarded me funds to begin a
study of how shell shape in snails varied geographically and
from low to high levels in the intertidal belt. The plan called
for work on the Atlantic and Pacific coasts of South America.
More by coincidence than by design, I would roughly retrace
the course of the *Beagle,* the vessel on which Charles Darwin

had made his famous voyage that led to his proposal of the theory of evolution. I would begin in northeastern Brazil and then travel by way of Rio de Janeiro to central Chile, Peru, southwestern Ecuador, the Galápagos Islands, Panama, and Curaçao. At each stop along the way, I arranged to stay at or near a marine biological research station.

All I needed now was a field assistant with a taste for adventure and a keen interest in natural history. Jim Porter was the obvious choice. As a Yale undergraduate, he had spent the previous summer chasing butterflies with Charles Remington and Ward Watt at the Rocky Mountain Biological laboratories in Colorado. Jim wanted to learn about everything and had a wholly optimistic outlook on life that made him quick with humor and laughter. His most distinctive talent was photography. Every scene was to him a painting, to be fully captured and enhanced on film. This gift made Jim a first-rate observer, one apt to notice even the most trifling details of an animal's behavior or a plant's leaf arrangement. My confidence that Jim would take to the sea as easily as he did to the high country of Colorado was amply justified. In later years, he would become a leading expert on the biology of coral reefs at the University of Georgia.

We set off from New York in June 1969. Recife, our first port of call, proved to be a logistical challenge. It was a large, chaotic city, buzzing with locally made Volkswagens as well as with banks that were unable to cash traveler's checks. With neither of us comfortable with Portuguese, I struggled to make myself understood with a combination of Spanish and French, but all efforts failed when we faced a taxi driver. "Mar Motel," we repeated, but to no avail. At last, Jim wrote the name of our friendly if seedy hotel down on the scrap of paper, but the illiterate driver pushed it away in disgust. We gave up, boarded a passing bus, and promptly got hopelessly lost. Eventually, we made ourselves understood to someone who pointed us in the right direction.

The shore fauna of Pernambuco, the state of which Recife is the capital, proved to be interestingly impoverished. There was only one minute local periwinkle species, and many of the intertidal snails familiar to me from the West Indies, including nerites, were absent. In their place was an array of uniquely Brazilian species, mostly of small size and all evidently tolerant of a thin veneer of sand on the inshore rocks.

"You should go to Fernando de Noronha," said Marc Kempf, a capable French marine biologist who had accumulated a broad knowledge of the Brazilian fauna during his tenure at the Instituto Oceanográfico in Recife. I had never heard of this volcanic archipelago. A tiny Atlantic outpost of some sixteen square kilometers, it lay about four hundred kilometers off the northeastern mainland coast of Brazil, just four degrees of latitude south of the equator. Darwin had sailed by it on the *Beagle* but had been unable to land. The Americans had now constructed an airstrip on the main island as part of a missile-tracking facility. The Brazilian army operated weekly flights on an ancient C-47, on which they ferried passengers gratis. For a sum that even an impecunious graduate student found nominal, the Department of Tourism and Public Relations (DETURP) put up visitors in a rambling guest house. Good meals of rice and fish could be had in the officers' mess.

Fernando de Noronha offered one of those defining experiences that shapes one's outlook out of all proportion to the time spent there. Though small, this isolated archipelago supports a distinctive fauna of endemic, or unique, shore molluscs that invited comparison with richer Caribbean assemblages. There is, for example, a single species of periwinkle. Rather than being restricted to a narrow zone on the shore, as most periwinkles are when several species share the available habitat, this one flourishes in large numbers over almost the whole width of the intertidal belt. It has the compromised shell morphology—a moderately high spire and a sculpture of moderately strong rounded beads—to match its catholic ecology. Upper-shore species in the Caribbean had higher spires and were usually more sharply ornamented, whereas the midshore specialists tended to have lower spires and to be smooth. My suspicion that this was an undescribed species related to periwinkles from Ascension Island, Saint Helena, and West Africa, was confirmed in 1982 when Bandel and Kadolsky named it as a new species, *Nodilittorina vermeiji*.

There is also a single nerite at Fernando de Noronha, which likewise combines morphological and ecological attributes of high-shore and midshore Caribbean specialists. After comparing it with closely related populations from the South Atlantic islands of Ascension and Trindade, I named the Fernando de Noronha form *Nerita ascensionis deturpensis* in 1970.

Most of the island had been deforested, so that only the wet western end could be described as lush. A huge population of goats, most of them outfitted with bells, ensured that the scrub would remain in a state of browser-induced siege. Their bells, together with the incessant crowing of roosters and the forlorn clanging of hammers emanating from a prison colony on the way to Ponta de Sao Antonio, make up some of the acoustical signature that resides in my memory.

We shared our spacious sleeping quarters with a population of uncommonly large rats. Once in the middle of the night I was awakened by one of the beasts heaving itself onto my bed. Back on the wooden floor, the giant soon encountered one of its colleagues, and the two engaged in a pitched battle over a Dutch peppermint that had fallen out of my pocket. A mauled length of tail was all that remained of the battle the next morning.

In Rio de Janeiro, we had our first encounter with grinding poverty and appalling pollution. We had looked forward to our stay at a small laboratory operated by the Instituto Oswaldo Cruz on Ilha Pinheiro, a small wooded island picturesquely situated in Guanabara Bay. A little rowboat would ferry us from the shore to the island. The dock lay at the foot of a filthy unpaved street that traversed one of Rio's infamous slums, or *favelas*. Barking dogs, screaming children, and the stench of burning garbage permeated the place. The island was little better. Its shore was lifeless, the polluted waters having killed even the hardiest mangroves. The acrid smoke from a huge dump, on which noisy bulldozers crawled day and night, mixed with the photochemical smogs of millions of gasoline and diesel-powered engines to produce a depressing brew of destruction and neglect.

The picturesque, well-kept laboratory at Montemar, on the coast of central Chile, offered a welcome relief. Fishermen brought their daily catch through an impossibly high surf to the sand beach in front of the lab. The adjacent rocky shores brimmed with the most extraordinary molluscs. There were five species of thick-shelled keyhole limpets, the foot even larger than the huge shell above; and on their backs the oval scars of *Scurria parasitica,* a small limpet that grazes plants from its host shell, interrupted the broad ribs. Still another limpet, the high-peaked *Scurria scurra,* excavated deep cavities in the stipes of its host kelp *Lessonia nigrescens*. The coiled snails, too, had large and conspicuously thick shells.

The theme of shell erosion illustrated by *Scurria parasitica* in Chile was carried to an even greater extreme on the mist-shrouded shores of Peru. Sponges, boring blue-green algae, and other tunneling creatures infested host shells so thoroughly that the older portions of many snail shells had simply disappeared, being replaced by rude plugs of lime. Only the most recently formed parts surrounding the outer lip still bore the sculpture typical of the species in question. On coasts as dramatically productive as that of central Peru, where the smell of guano is as pervasive as the dank fog, shell destruction by eroding creatures was clearly an important factor to the molluscs that live there.

Every evolutionary biologist yearns to set foot on the Galápagos, and I was no exception. Darwin had trumpeted the uniqueness of the terrestrial fauna, which differs slightly from island to island and which provides compelling evidence for the local formation and adaptation of species from ancestral invading stocks. The marine biota is no less thought-provoking. The islands support a rich tropical fauna of molluscs, many of them endemic and most of them well hidden under stones and in pools from the equatorial sun that heats the basalt on the shore to stovelike temperatures. Yet there are strangers here, intruders from the temperate south whose ancestors must have followed the cold Humboldt Current from South America. How else can one account for endemic black gulls, flightless cormorants, penguins, and sea lions living side by side with tropical corals, nerites, and parrotfish? Where else can one hear marine iguanas sneezing out salt water while they warm themselves on rocks in the sunshine before plunging into the sea for another meal of seaweeds? Why should only the Galápagos have provided this opportunity for a land-derived lizard to become an important herbivore in a marine ecosystem?

The Galápagos visit very nearly proved to be our undoing. As our Ecuadoran military DC-4 took off from the overly long runway at Isla Baltra for the return flight to Guayaquil, the propellers on one side of the aircraft seemed to be rotating more slowly than on the other, with the result that the plane vibrated intensely every few seconds. With a sudden jarring gasp, the periodic shudders ceased, and one of the propellers suddenly stopped turning. Having been in the air for only an hour, we could look forward to two more hours until we would reach the coast. Would we ever get there? A rough landing or crash seemed inevitable. To my horror, the space beneath my seat where the trusted life vest should be was empty. Darkness was coming on, and our chances of being rescued were in any case small in this deserted part of the Pacific. One of the passengers, a Swiss engineer, went forward to consult with the frustrated pilot. With his guidance, the plane stabilized and was able to limp back to the military airfield at Guayaquil, but not before Jim and I had time to reflect on the fragility and precariousness of life. Several years had to elapse before I lost my fear of flying.

After Fernando de Noronha and the Galápagos, almost any place would look pale and ordinary, but Panama was hardly an anticlimax. Few locations can match its remarkable geography and diversity. On a single day, one can sample the abundant marine life on the sun-baked shores of the Bay of Panama in the eastern Pacific and wade on the reef flats of the Caribbean coast not more than fifty kilometers away. The contrast between these two worlds is stunning. On the Pacific side, there is a huge tidal range of four meters or more, and much of the rock is of igneous origin. Expansive flat sandy beaches stretch for miles along the coast, and everywhere there is a profusion of marine life dominated by molluscs, barnacles, stinging hydroids, and encrusting invertebrates of all sorts. The Caribbean side presents a wholly different picture. The tide rises and falls no more than thirty centimeters, and the steady trade wind can make even this fluctuation highly unpredictable. Most of the rock is limestone, formed by the corals that grow on the reef just offshore. A large and conspicuous cover of seaweeds blankets the reef flat, whereas on the Pacific

side such an algal turf is virtually absent. In keeping with the
low availability of food in the water over the reef, molluscs are
not overwhelmingly common as they are on Panama's Pacific
coast, where upwelling of deep nutrient-rich waters during the
dry season from December to April bathes the shore in a
plankton-rich soup.

Panama fired the imagination. Here were two drastically
different marine biotas, separated by a sliver of land for not
more than a few million years, perhaps soon to be reunited by
a sea-level canal that could replace the present freshwater passage. How and when did the contrasts come about? How
would the two coasts be affected if species were free to disperse between them? I was not ready to tackle such questions,
but they were firmly planted in my mind for later harvest.

The summer's results confirmed my earlier guesses about
snail shape on the high shore. The succession of species from
the midshore to the upper reaches of marine influence was architecturally similar to the succession of species at high intertidal levels from the middle latitudes to the tropics. Did these
patterns in shell shape reflect common underlying causes?
How do snails living in a regime of heat, drought, and a
scarcity of food cope with the stresses of their environment? I
needed to know what these stresses were, how they varied
from place to place, and to what extent the shell played a role
in enabling snails to eke out a living in the hostile habitats
above the tideline.

Because the Caribbean region supports an unusually diverse fauna of high-shore snails—eight species of periwinkle,
four nerites, and half a dozen other species—it seemed sensible to begin answering these questions at the Discovery Bay
Marine Laboratory, on the north coast of Jamaica. Thomas F.
Goreau, the world's leading reef biologist of the day, had established the lab several years earlier as a base for his diving

operations. The lab was housed in a little concrete shack on the east side of the bay. I had paid a brief visit in January 1969, but my stay in early 1970 would be for two and a half months.

The lab was ideally situated for the trail-blazing research on coral-reef environments that was being carried out in Jamaica, much of it by fellow Yale graduate students. Immediately offshore, a luxuriant reef stretched for miles, its seaward face descending steeply to a depth far in excess of one hundred meters. Although I did not dive, reef-building corals grew in water shallow enough for me to wade among them and gain an appreciation of reef structure. One such place was Fort Point. Exposed to the full force of the surf, it marks the northeast entrance to Discovery Bay. Above the tideline, Fort Point consists of sharp and craggy limestone, referred to locally as ironshore, a landscape of pinnacles and flat-bottomed round pools surrounded by razor-sharp points and edges. At its seaward margin, this terrain drops vertically as a two-meter-high cliff to a narrow ledge, which is accessible only during periods of extreme calm and even then only at one's peril. The ledge, slick with growths of stony corallines and larger fleshier seaweeds, is fringed by a dense forest of staghorn corals and waving sea whips, marking the uppermost reaches of the reef below. More than three dozen species of mollusc live here, including a host of keyhole limpets, chitons, and snails ranging in size from a few millimeters to more than ten centimeters.

Tom Goreau had created an intensely creative atmosphere at the lab. Irascible, impatient, and always interesting, he possessed an unparalleled vocabulary of profanities. His recklessness with language was surpassed only by his daredevil diving. Frequently, he descended to a depth of one hundred meters on the forereef, where he spent the few available seconds exploring dark caves. In these he discovered puzzling growths to which he referred as beasts. Goreau had sent specimens to many of the world's invertebrate zoologists, most of whom were as baffled as he was. Only Hartman recognized them for what they were: slow-growing, heavily calcified sponges, cryptically living remnants of a once diverse group that dominated reefs of the distant past. With these sponges lived other relics such as small brachiopods, as well as hundreds of other invertebrates. Goreau, Hartman, and Jeremy Jackson published a paper on this community of relics in *Science* in 1971.

Thanks to Goreau and the students who gathered around him, the reefs of north Jamaica became the most thoroughly studied in the world. Judy Lang, who worked with Hartman and Hutchinson, had discovered a hierarchy of aggression among reef corals, the most dominant species being capable of digesting all the others and the most subordinate species being susceptible to aggression by those of higher rank. Henry Reiswig, another student of Hartman's, was ingeniously measuring the rates at which sponges drove water through their elaborate filtration system to extract food particles whose size and quantity he was also able to estimate. Robert Kinzie's work on sea whips, and David Meyer's on sea lilies, threw light on other important reef dwellers. Jeremy Jackson concentrated his labors on the population dynamics of bivalved molluscs in the grassbeds of the backreef. Joining this Yale contingent were David Barnes from England and Philip Duston from the State University of New York at Stony Brook, both working with corals. Colorful, humorous, dedicated, and immensely helpful to me, this was a wonderful collection of individuals working on shoestring budgets and with the most basic equipment in an atmosphere brimming with opportunity and discovery.

The dark cloud that hung over Discovery Bay was Goreau's grave illness. Tom had long suffered from ulcers, but now the stomach pains were becoming intolerable, and he groaned and grunted uncontrollably. He had labored long and hard to see to it that a new lab, some fifteen times the size of the old one, would rise on the western side of the bay, opposite Fort Point. In the presence of Princess Alice, a granddaughter of Britain's Queen Victoria, the facility was officially opened on March 23, 1970. Less than a month later, Goreau was dead of cancer, a tragic loss for the world of marine biology.

The shores of Discovery Bay are almost all limestone. So are those on Curaçao, where I spent two weeks in April 1970. Limestone is less hostile to the snails of the upper shore than are rocks of volcanic origin, which not only become hotter but are also less easily penetrated by the rock-boring seaweeds that form the bulk of the diet for the snails. If I wanted to observe snails under the most extreme conditions of heat and desiccation, I would have to visit volcanic shores, especially those in the Pacific, where I had made the original observations prompting my interest in the higher intertidal habitats. I therefore began planning an ambitious round-the-world trip that would take me to Hawaii, Guam, Palau, the Philippines, Singapore, and the Netherlands. At each of the tropical sites, I would hire a local field assistant. It was a great idea but, as always, I needed money. From Jamaica, I applied for funds from the Geological Society of America, which generously awarded me $1000, and the Organization of Tropical Studies (OTS). The latter organization, which almost single-handedly trained a generation of tropical biologists at its facilities in Costa Rica, advertised small research grants for work anywhere in the tropics.

Back in New Haven, my anxiety about funding was compounded by the social upheavals of the spring of 1970. Against the backdrop of mounting opposition to the Vietnam War, a student strike in support of the Black Panthers was in progress.

All the buildings were in a state of high alert against sabotage. Ingles Rink, half a block down the street from my dingy room on Prospect Street, was bombed in early May. On the whole I found myself unsympathetic to the politics of the protesters. To be sure, I was as unhappy with the interventions in Southeast Asia as were most of my peers, but I disliked the pressure tactics of strikes, demonstrations, and binding resolutions. In any case, they convinced me that I wanted to be in the field, far from the political ferment in New Haven and from the inevitable tensions that would surface during the hot summer in that racially divided city.

The news from OTS was not encouraging. My proposal was very good, they wrote, but OTS was funding only work done in Central America. In desperation, I protested, pointing out the inconsistency between this policy and the one proclaimed in brochures describing the research grants. I took the phrase "anywhere in the tropics" to include the tropical western Pacific. To my amazement, officials at OTS agreed and awarded me the full sum of $1850 that I had asked for.

I began my trip with a reacquaintance with what was euphemistically called Apartment 1, the little barracks on Coconut Island where I again slept in the company of cockroaches and toads. I took my meals with two other visitors who shared one of the nicer apartments at the institute. Ariel Roth and his student Conrad Clausen were coral physiologists from Loma Linda University. As Seventh Day Adventists, they kept to a strictly vegetarian diet, which they ingeniously enlivened with an array of tasty dishes made from grains, beans, and vegetables. These were not the typical fundamentalist Christians whose views on evolution and religious matters amount to doctrine ladled out of a can. Roth and Clausen understood what they thought and why they thought it, and I admired them for their willingness to discuss our differences so dispassionately and intelligently. They were also notably generous. On Saturdays they invited me to the church in Kaneohe, where an easygoing and undogmatic service was followed by a sumptuous potluck meal. Perhaps the soft Hawaiian climate and the island spirit of generosity and community took the edge off the zeal with which Christians so often push their beliefs on others. Where before I had perceived mostly evil and ignorance, I began to appreciate that people with strong religious convictions radically different from my own atheist outlook could be as thoughtful and benevolent as anyone else.

On Guam, Lu Eldredge had made arrangements for me to work with Sebastian Ongesii in Palau. Sebastian, his wife Prisca, and their two young daughters were planning to visit Prisca's brother Thi, and I could come along to stay in Thi's house in Ngerchemai, one of the villages on the main island of Koror. The floor, raised some thirty centimeters above the ground, was made of beautifully smooth ifil, a fine-grained dark wood that sinks in water. For almost three weeks, I immersed myself in a rich Pacific culture. Only Sebastian and Thi spoke English. The rest of the household, which included Prisca's ancient grandmother as well as a dozen or so children, conversed in Palauan. To the untrained ear, this remarkable language is an unbroken succession of consonants, mostly uttered far back in the mouth and contrasting sharply with the vowel-rich languages of Polynesia. I was treated to delicious meals of fresh fish, rice, taro, and tapioca, often with taro leaves cooked in coconut milk as a vegetable. Japanese noodle soup, sometimes with a side dish of squash in coconut milk, served as breakfast. Giant clams, their slightly sweet meat made even more so by coconut milk, occasionally substituted for the fish. The taro root had the consistency of potato, but its flavor was mildly like beer.

Palau's fauna was richer and more spectacular than any I had yet seen. No less than a dozen nerite species lived in Palau's shores; some species occupied only mangrove roots, others only limestone, and one occurred mainly on volcanic rocks. Beds of turtlegrass teemed with large scorpion shells and hammer oysters, and hand-sized sea stars with huge dorsal tubercles lay motionless on the meadow's surface. On the barrier reef surrounding the archipelago and its great lagoon were large, incredibly thick-shelled species of *Vasum* and *Thais* encrusted with huge yet inconspicuous flat barnacles neatly hidden among the knobs and spines of the shell surface. This was tropical exuberance, the unfettered architectural diversity that I had dreamed about as a boy.

My plans in the Philippines were much less secure. On the recommendation of Roy Tsuda, a young seaweed expert who had come to Guam in 1969, I wrote to Gavino Trono, who worked on seaweeds at the University of the Philippines. By the time of my flight from Guam to Manila, however, I had heard nothing from him. The chaotic airport was not the kind of place I wished to be stranded, but I need not have worried. With great relief, I found Dr. Trono waiting for me, ready to drive me to my comfortable lodgings at the university's campus in Quezon City.

If I had hopes that Rio de Janeiro's deplorable condition was a unique aberration for a large tropical city, they were decisively dashed by Manila. Millions of vehicles, powered by dirty diesel and gasoline engines, choked the air with a concoction of hydrocarbons and ozone, which together with the smoke of the cooking fires and garbage made for a nauseating smog worse than anything I could imagine. In the midst of it all lived millions of destitute people, for whom this devastated place was in some mysterious way superior to the country life many of them had forsaken. My graduate student stipend seemed grotesquely generous, even to Trono and his fellow professors at the university. For them, hiring a driver and jeepny to go to Calatagan or Nasugbu was out of the question. I gladly paid Trono's way to Puerto Galera on the island of Mindoro, though Trono might well have been sorry. The ride across storm-tossed Mindoro Strait in an outrigger banca was so bouncy that the three-man crew repeatedly crossed themselves for salvation.

Singapore, my last tropical stop, introduced me to spectacular cultural and biological diversity. Three languages— English, Malay, and Chinese—were in common use there, and my assistant Seow spoke each of them fluently. He introduced me to the full range of Asian cuisines, including such uncompromisingly hot and tasty dishes as *nasi brani* (chicken and rice), as well as fried noodles with *Anadara granosa,* a local mangrove clam.

The most impressive biological diversity on Singapore's shores was to be found in its mangrove forests. Some twenty-five species of salt-tolerant trees and shrubs grow in a dense tangle of branches, trunks, and prop roots on quiet sandy and muddy shores. On other mangrove shores I had visited— Guam, Palau, the Philippines, Ecuador, Brazil, Jamaica, and West Africa—only one or two species of periwinkle clung to the branches of at most five species of mangrove trees, but in Singapore I could find six periwinkle species, some living low among the barnacles and oysters, others high above the water on branches and leaves. Where other mangrove swamps would have supported at most one nerite, Singapore had three species. Half a dozen species of horn snail hung by mucus threads from branches or crawled on mud so soft and so deep that I sank into it up to my thighs as we oozed our way to the seaward fringe of the forest. Perhaps strangest of all were the limpetlike jingle shells, small clams that crawl on tree leaves and branches by means of a prehensile foot. On the lower parts of the trees, every surface was festooned with razor-sharp oysters and large barnacles, among which crawled a pleasing diversity of predatory snails.

With Singapore situated halfway around the world from New Haven, it was as economical to return via Europe as it was to fly back across the Pacific. I wanted not only to see Arie but also to make my reacquaintance with Henry Coomans, who was now installed at the Zoological Museum in Amsterdam. The substantial collection of Indonesian molluscs at that venerable institution would perhaps enable me to put some names on the species with which I had worked in the tropical Pacific.

After three and a half months in the wet tropics, the unmistakably musty odor of mildew had seeped into all of my belongings. Even if I hardly noticed, my brother Arie and his new wife Hanny upheld the Dutch tradition of cleanliness and were quite unaccustomed to such a nasal assault. When I opened my bags in their spotless flat in Zeist, they staggered back in horror. Bags of shells from Guam reef flats, Philippine beaches, and Singapore mangroves were quarantined to the balcony while all my clothes were hastily dumped in the washing machine.

It felt comfortably familiar to be back in the Netherlands for a month's stay. Hanny's delicious Dutch cooking, the voices on the radio, the smell of pine woods in nearby Huis ter Heide brought back floods of boyhood memories. But I had changed. The tropics had left me with more than just a few moldly suitcases and a stack of data for my dissertations. Holland was built on such a small scale; it was so busy, yet so orderly and neat, its woods manicured, its population healthy and prosperous. I had grown accustomed to spicy foods, to steamy nights without bed covers, and to a constant chorus of crickets and geckoes. I had been brought face to face with poverty, degradation, overpopulation, and perhaps most of all the powerful attraction of truly wild places.

Having a quiet sun-drenched room in which to work, I set about analyzing the field data and writing the dissertation. It was now clear that periwinkles and nerites deal with the exigencies of high-shore life quite differently. When exposed to the air, periwinkles withdraw the foot into the shell behind the operculum, which makes a nearly hermetic seal from the outside. A thin film of mucus holds the animal firmly in place on the rock, branch, or grass blade. In this inert state, the periwinkle loses almost no water and maintains minimal contact with the often hot surface to which it clings. The tall spire keeps the surface area that is directly exposed to solar radiation low, whereas the beads and ribs on the shell's exterior surface reflect heat. Nerites, by contrast, rely on water evaporation for cooling. The interior of the shell has lost its spiral structure, being transformed into a spacious water reservoir. The spherical form of the species at the highest shore levels combines a relatively large reservoir with a somewhat reduced contact area between the snail's foot and the underlying rock. Should the nerite be dislodged, it too can withdraw behind a tightly fitting heavy door and survive for days without water. At lower shore levels and higher latitudes, the problems of heat and desiccation are reduced, and specializations to insulate the soft parts from the weather outside are less evident.

The dissertation left many questions unanswered and many assumptions insufficiently documented. Hindsight clearly shows that I should have complemented observations of tissue and environmental temperatures with experiments. It would have been instructive to see if a modification of shell sculpture affects the thermal relationship between a periwinkle and its environment. Most important, I should have ascertained whether the physical rigors of the upper shore are sufficient to cause selection in favor of beaded sculpture, high spires, and a large spherical reservoir. Its obvious faults notwithstanding, I planned to write the dissertation as four journal articles, an approach that Yale fortunately encouraged. Then as now, I saw no merit in submitting data in a unified dissertation that I would not also submit for publication. If the details are important to the arguments presented, they should be published; if they are not important, there is no good reason to present them in any format. Mid-October found me back in New Haven. Now came the greatest challenges of all.

Chapter 8

SKATING ON ICE

Much as I appreciated the freedom of being a student and the generosity of the stipends enabling me to exercise this freedom, I was at all times keenly conscious of my financial dependence on others. I could not claim success or equality in my chosen profession until I earned a legitimate income. I resolved to complete my Ph.D. as quickly as possible and to find a teaching post. I wanted to be at a university where I would be free to conduct research while teaching bright, motivated students about evolution.

Against the advice of several friends, I decided not to seek an interim postdoctoral research position. True, such a position would have allowed me to do research with minimal interference, but taking on such a temporary appointment would only have postponed the task of securing permanent employment. The misgivings of my counselors at the Commission for the Blind gnawed at me with increasing urgency. I might have been self-confident or arrogant enough to dismiss their doubts as the uninformed opinions of overly cautious people who were not quite willing to place complete faith in the competence of blind people, but I also knew that their worries were not unfounded. In 1971, the golden age of academic employment was coming to an end. So was the most flagrant manifestation of the old-boy network, in which famous senior professors could place their favorite intellectual offspring more or less where they pleased. Moreover, whereas some university departments might be willing to gamble on a blind graduate student, most were unquestionably more reticent to hire a blind person as a member of the faculty. They could always fire a new professor if that person failed to measure up, but the process is painful and time-consuming, and with the cap on enrollments that many universities were now forced to accept, there was no guarantee that the unsuccessful recruit could be replaced by someone more suitable. Any inexperienced young Ph.D. would be a risk; to appoint a novice who was also blind might seem suicidal to a cautious department. And in my case there was further cause to worry. I had been in graduate school only three years. Not only did I not have any postdoctoral research experience, I had never even been a teaching assistant or had any teaching credentials of any kind. All that anyone had to go on were letters from my mentors and a small collection of published or in-press papers that by the end of 1970 amounted to five items.

Without the unqualified support of influential people, I would go nowhere. Hutchinson, whose gentle and optimistic guidance set many an illustrious career in motion, still wielded great influence in all the biological disciplines related to evolution and ecology. His support would be essential. He had a long history of looking after his protégés, who included many women as well as a few intellectual mavericks traditionally shunned by the academic establishment.

That Hutchinson had been actively pleading my case became clear when a letter arrived from John Corliss. As the new chairman of the zoology department at the University of Maryland at College Park, Corliss was anxious to build a group interested in ecology and marine biology. A world authority on ciliated protozoans, Corliss hoped to bring together a faculty that would concentrate its efforts in Chesapeake Bay, where the university already owned one research lab and was planning to found another. Corliss wondered if I might like to apply. Chesapeake Bay, being a low-salinity estuary, has a fauna that is impoverished even by the low standards of the marine areas of eastern North America. The northwestern Atlantic Ocean supports one of the poorest marine temperate biotas in the world not only because of its notoriously variable marine climate, but also by virtue of a catastrophic series of extinctions that wiped out 70 to 80 percent of species living two to three million years ago. For me, therefore, the prospect of working in Chesapeake Bay rather than some outrageously diverse tropical location was unappealing. Nevertheless, why burn one's bridges at this early stage? After some hesitation and with considerable reluctance, I submitted the necessary materials with my application. At about the same time, I applied to Johns Hopkins University and to a few other institutions, but I quite foolishly passed up many others that Hutchinson and Hartman commended to my attention.

Employment was by no means my only concern about the future. After returning from the Netherlands, I perceived more clearly than ever the potential pleasures of marriage. My brother had done very well by Hanny, and many of my peers were also either engaged or happily married. How or whether I could ever make it to that state was quite uncertain, given the scant attention I had given my social life. To say that I was shy and socially inept would be a tremendous understatement. I never attended dances in high school, and the few parties to which I was invited at Princeton reinforced my conviction that large gatherings were ill-suited for meeting the kind of reflective woman I might like.

I had had only one previous tentative brush with romance. Within two weeks of my arrival in New Haven in 1968, I met Edith Zipser, a third-year graduate student in Dick Goldsby's molecular biology lab. We dated occasionally, but as the year wore on and I was preparing to go to South America with Jim Porter, we drifted apart. Unflattering though it is to admit it, I instinctively and unthinkingly placed work ahead of all else.

Now, two years later in the fall of 1970, I weighed the options. There were, to be sure, some very appealing women—intelligent, interesting, well-spoken, and with a pleasantly feminine voice—but some rather badly bungled investigation on my part quickly revealed that most were already spoken for. When approached with an invitation to dinner, one responded with surprise over my ignorance that she was engaged to be married the next week.

Jim Porter unwittingly set the stage for a most satisfactory solution. He had been participating in a lively discussion group led by Hutchinson about the relationship between science and culture. With Kenneth Clark's book and television series *Civilisation* as a point of departure, the group considered how science and the arts provide the necessary ingredients of culture. Hutchinson, that quintessentially Renaissance gentleman, was as much at home in medieval art as he was in English literature and the arcana of mathematical modeling in ecology. Even in casual conversation, he drew heavily on his breathtaking knowledge. He attracted a devoted group of the most intellectual students around him, a group that I was foolishly too preoccupied to join, but from which both Jim and Edith derived much pleasure. After one of the sessions, Jim remarked that I was looking for people to read to me. Remembering the location of my tiny office in the geology building, Edith came by for a chat and offered to do a little reading.

As had been the case at Princeton, the Commission for the Blind provided a monthly sum with which I could pay fellow students to read. Cooperative librarians allowed me to take books and journals out for short-term loans so that I could have them read to me in my office without disturbing others. This reading was not only an academic necessity but also one of the few opportunities I had for getting to know people without the discomfort and awkwardness of larger social settings. I was therefore more than a little delighted when Edith appeared at my door. Not only was she an extraordinarily fast and accurate reader, but I was immediately taken with her clear bell-like voice, obvious thoughtfulness, and the genuine curiosity she displayed about the shells that were crowding me and my books out of the cramped space of my office.

During the next several weeks, I set Edith to reading parts of D'Arcy Thompson's *On Growth and Form*. Perhaps she would find his literary style and profound knowledge of the classics appealing. I frequently called her in Dean Rupp's lab, where she was now busily engaged in research on DNA repair in mammalian cells. After enjoying dinner together, I would offer Edith some of the jasmine tea I had brought back from Singapore.

Liking was soon replaced by love. Never before had I experienced anything like it. Love was powerful, a whole new dimension of life and a state of mind that seemed to brighten everything around me. Not long after our reacquaintance, my love for Edith had become so deep that I felt ready to spring the news of my involvement with her to my parents. Edith was, I explained, the perfect woman for me. Of course, she had a sharp mind and broad interests, and she possessed a serious disposition. Like me, she appreciated classical music, good food, long walks, and the works of nature. And then there was her hair, too good to be true. In contrast to my very ordinary straight brown strands, hers were excitingly curly. Tight spirals and gentle waves made for an exquisitely woolly blanket, springy and warm to the touch.

Any prospect of serious romance had to yield to the grim reality of my economic predicament. Marriage was out of the question unless I could find employment. It was therefore with great relief that I learned in December that I would be granted interviews at both Johns Hopkins and the University of Maryland. It was now up to me to persuade the discerning faculties of these two institutions that I would make a useful and interesting junior colleague.

Of the two jobs, I wanted the one at Johns Hopkins more. Steve Stanley, my host, was someone I had come to admire. Like me, he had worked at Princeton with Alfred Fischer and then moved on to Yale as a graduate student. There, he distinguished himself by writing one of the finest dissertations in memory. He carefully and elegantly analyzed how the shapes of clam shells correspond to the mode of life of the animal, and he used simple principles of mechanics to show how form and function were interrelated. To a significant extent, his work served as a model for my own much less complete study of snail shells. Now Stanley was at Johns Hopkins, anxious to hire another paleontologist. Here was a genuinely exciting opportunity.

The two interviews were scheduled back to back, with the one at Johns Hopkins coming first. Stanley met me at the train station in Baltimore, and we hit it off immediately. As I remember it, the interview proceeded smoothly and uneventfully. None of Stanley's geological colleagues raised any overt objections to my blindness, and the lectures summarizing my research went well. The idea of working in the middle of

Baltimore did not thrill me, but I was impressed with the department and thought highly of the university as a whole.

I was, however, not the only candidate. Jeremy Jackson, a fellow graduate student in geology at Yale, was the other leading contender. He was a worthy competitor. I had liked him as soon as we met in the fall of 1968, when he volunteered to do some reading. In Jamaica, he helped me in many important ways with my snail work. By the time Jeremy interviewed at Hopkins, he was in the fortunate position of having two highly worthwhile research projects to describe. Besides his thesis on grassbed clams, Jeremy could point to his work with Goreau and Hartman on the Paleozoic-like deep reef communities. Moreover, he had a warm, polished manner and a deep clear voice that produced a long sonorous laugh.

With the Hopkins interview behind me, my attention turned to College Park. John Corliss met me in Baltimore and began at once with boundless enthusiasm and animated manner to outline his plan for the rejuvenation of his department. As we munched on stuffed oysters in one of Baltimore's fine seafood restaurants, Corliss talked of the new zoology building, which would be shared with psychology. I felt at ease with him. He had a breezy, good-humored style and was always ready with anecdotes about French protozoologists or about his youth in Vermont. By the time the formal interview began after breakfast the next morning in College Park, I felt relaxed enough to venture a few puns.

Nevertheless, the interview began inauspiciously. A sore throat that had irritated and fatigued me at Johns Hopkins had now turned into a full-fledged cold. Even worse, the weather turned to freezing rain, covering every surface with a treacherous coating of ice that challenged my self-confidence.

As at Johns Hopkins, the interview itself proceeded without incident. I made the round of the professors who would cast votes to choose their next colleague. There were the usual polite questions. How would I lecture? How would I grade exams and papers? Not having done any of these things, I could only guess at the appropriate answers by recalling what my own mentors had done so successfully. I would spell out words, have things written or drawn on the board, hand out summaries, and bring specimens for demonstration in class. Like other professors, I would expect to have teaching assistants who could help read students' work. The answers seemed to satisfy my inquisitors, none of whom expressed any reservations.

What lecturing skills I possessed were about to be displayed for all to observe at the presentation of my research before the assembled members of the department. Somewhat to my surprise, the audience included Joseph Rosewater, curator of molluscs at the Smithsonian's Museum of Natural History. As the world's expert on the classification of periwinkles, he was someone whose opinion would carry substantial weight. I asked someone to write a few equations on the blackboard, and I had specimens ready to be passed around the room. As I explained how the various snails of the upper rocky shore dealt with problems of water loss, overheating, and heat loss, I tried to show how equations describing heat transfer could be employed to understand the adaptive possibilities that were available to shell-bearing snails and to assess the effectiveness of the snails whose shells I had brought along. Rosewater and the others seemed genuinely intrigued, and the questions they asked at the close of my formal presentation indicated that I had been successful in communicating the main points of the argument.

The final event of the day took place at the home of Douglas H. Morse. It was traditional for candidates to engage in informal conversation and to answer lingering questions about the seminar at this so-called brain pick. Morse, who is today the Herman C. Bumpus Professor at Brown University, was a young New Englander working on the niches and food habits of warblers, the birds on which Robert MacArthur had earlier based his classic thesis. He was a seasoned naturalist who cared little for the inflated theoretical underpinnings of ecology and for the pretenses that often came with the theories. I liked his expressive deep loud voice with the strong Maine accent, as well as his plainspokenness and thorough knowledge. He was clearly a man of substance who would make a fine colleague, I thought to myself as questions and fatigue battled it out in my congested head that icy evening.

The next several weeks were anxious ones. What would I do if neither interview was successful? Could I somehow find money to remain at Yale doing postdoctoral work, so that I could stay close to Edith? Were there other positions for which I might still apply? Were the curmudgeons at the commission right after all?

Without a degree, however, all these worries would be moot. It was time to knuckle down to the last stages of writing my doctoral dissertation. I wanted, moreover, to make a few additional measurements of snail and environmental temperatures, especially on West Indian periwinkles living on volcanic shores. The Windward Islands of the eastern Caribbean would be ideal. With some funds that Hutchinson had at his disposal, I traveled with Jim Porter to several eastern Caribbean islands. He took his fiancée, Karen Glauss, a graduate student in limnology working with Hutchinson and John Brooks, and I persuaded Edith to come. If Edith and I were to become seriously involved, I reasoned, I should discover whether she enjoyed fieldwork.

For two glorious weeks, we traveled from St. Maarten to Saba and Guadeloupe. With Edith and Jim's enthusiastic

help, I obtained my temperature measurements and made observations on the natural history of other shore snails. Edith,
not surprisingly, turned out to be a very competent field assistant, to say nothing of being a delightful companion. The
four of us endured a filthy hotel in Pointe à Pitre, the capital
of Guadeloupe, and joked about the large number of anthills
in the Antilles. We climbed the mountain on Saba and, on
our last day, visited the Soufrière volcano on Guadeloupe.

There had been some very considerable discussion about
whether we should take the extra day to visit this impressive
site. I was increasingly concerned about getting back to New
Haven. Much of my dissertation remained to be written, and
the deadline for submission of the finished product in time for
a June degree was only a month away. Fortunately, my com-
panions prevailed on me to stay. The mountain left an indelible
impression. Its slopes were richly clothed in a tangle of vines
and epiphyte that contrasted with the bleak and barren summit, where fumaroles expiring noxious sulfur made it all too
easy to imagine how notions of the underworld of fire and
brimstone developed in the Mediterranean world.

Almost every day during the trip, I called Hartman to ask
if he had received word about my job interviews. Each day the
answer was the same: no, he had heard nothing.

Word finally came after our return to New Haven. John
Corliss called to offer me a position as instructor. I was to let
him know by the end of February whether I would accept.

This offer presented me with a dilemma. Johns Hopkins,
whose position I frankly wanted more, remained silent. The
more vexing problem was how I should respond to the somewhat unorthodox terms of the Maryland offer. The position of
instructor does not lead to tenure and is usually a temporary
appointment. Two other candidates looking for their first academic jobs had been offered assistant professorships and were

therefore on track for eventual promotion to the tenured ranks. Corliss offered me the salary that a beginning assistant professor would earn and assured me that I would have a very good shot at joining the ranks of the tenurable. This was, he said, the best he could do for now, but he would work hard to put things right at the earliest opportunity.

Despite his assurances, I remained skeptical. Was I, as a blind applicant, not being asked to meet a higher standard than were others of comparable experience? Was this not unfair? On the other hand, I understood deeply that Corliss and the rest of the zoology department were taking an uncommon risk and that they were doing what they could to give me an opportunity without taking what they perceived as a potentially disastrous gamble.

As the weeks of February wore on, I was mostly occupied with writing. For the first time in my life, I was staying up until midnight as I worked ceaselessly on the last of the four chapters making up the dissertation. This preoccupation, exacerbated by the uncertainties about my job future, did little to endear me to Edith. I would come to Edith's apartment on Trumbull Street, gulp down a meal, and then promptly disappear without so much as laying a hand on the dirty dishes.

Soon, however, the day came for my dissertation defense, and the mad rush was over. Committee members posed polite questions, joked and laughed while I waited anxiously outside for several long minutes, and then pronounced me Doctor of Philosophy.

The job dilemma, too, resolved itself. Jeremy Jackson, it turned out, would get the Hopkins job. My disappointment was blunted by the great admiration and respect I felt for him. Already he was a formidable scientist, someone whose vigor and scientific breadth would propel him as a big player to a position of importance. In the years since, Jackson has indeed become a prominent force in the study of clonal marine invertebrates, coral-reef biology, and the faunas and history of marine tropical America.

With a certain reluctance, I accepted Maryland's offer at the end of February. The move to College Park would entail prolonged absences from Edith. It was therefore high time to forge my relationship with her. Even in New Haven, the spring is a forgiving season, when the smell of freshly green grass and the songs of robins, sparrows, and blackbirds awaken and nurture the romantic spirit.

My parents must have been greatly relieved when their shy and socially awkward son finally managed to find a serious girlfriend. When they first met Edith shortly after my dissertation defense, their delight was palpable. Many years earlier, I had half-jokingly confided to my mother that I would marry a bright, articulate, reflective, and almost certainly Jewish woman. Here she was in person.

A month later, it was my turn to meet Edith's parents in Westbury, Long Island. Her father, Samuel Zipser, was the highly successful president of a small advertising agency in New York, where Edith's mother Sylvia also worked. In their tidy split-level house, the Zipsers were gracious hosts, outdoing themselves with delicious food as a large group gathered for the Seder festivities before Passover. Yet I sensed their misgivings. Not only was I blind, but I was not Jewish. Yes, I had found a job, but could I really hold my own in the rough-and-tumble world of competitive work? Nonetheless, they hid their skepticism well and never expressed any of their doubts to me. On the contrary, their generosity and support knew no bounds as our courtship deepened.

Much as I dreaded the day when I would have to leave Edith behind in New Haven, I was anxious to plan for my new life at Maryland. At the top of my agenda was to find a pleasant place to live within walking distance of my campus office. At Yale, I had grown accustomed to the independence and convenience of traveling with the aid of a white cane. As a fast walker with quick reactions, I could manage with the very short canes of those days. Mine came up only to the top of my breastbone, so that I had to hold my right arm in a perpetually extended position to gain enough warning of impending obstacles in my path as I swung and tapped the cane from left to right and then back again in front of me while walking. Today's much longer and lighter-weight canes have made walking far safer and more relaxing, but even the inadequate canes were to me preferable to a guide dog. In the far-flung places I frequented, a dog would be constraining and utterly impractical. Cane in hand, I traveled to College Park in May to acquaint myself with the new environment. With the generous help of Doug Morse's energetic wife Elsie, I quickly found a small apartment in College Park, some fifteen minutes' walk across the campus to my office in the new Zoology-Psychology Building. The route, which was immensely complicated with innumerable turns, steps, and odd angles, took me some time to master, but the independence was worth the effort.

There was, of course, more to independence than commuting by foot. I also learned how to cook. With recipes and hints from Edith, my mother, and my brother's wife Hanny, I quickly discovered the satisfaction of preparing my own meals. Stews, chicken dishes, soups, Dutch nutmeg-laced meatballs, eggs, large salads, steamed vegetables, potatoes, and rice formed the core of my evening meals. I learned how to make tasty Dutch puddings and butter cake. Although my efforts never resulted in dishes the equal of those of the women from whom I inherited the recipes, I welcomed cooking as a diversion from the day's work, and the smell of my own creations steaming, frying, or baking added an element of domesticity that I remembered with such fondness from home and that I longed to recreate with Edith.

The rehabilitation counselors at the Commission for the Blind could finally close their books on my case. I was educated and employed, ready to repay the tax investment that had so generously been lavished on my training. For me, however, this was only the beginning. Promises and opportunity now had to be fashioned into contributions and accomplishment.

Chapter 9

CHOICES

There was no mystery about what was expected of a young faculty member seeking the lifelong security of tenure at a university. No graduate student at Yale, where the high-stakes tenure drama was played out in its unfettered ferocity and ruthlessness, could fail to understand that research was a necessary key to success. It couldn't be just any research; it had to be important, path-breaking, modern, and well-funded, published as a steady output of papers in the most prestigious scientific journals. By these measures, the molecular biologists held a decided advantage. Although they concentrated on a mere handful of model organisms—yeast, bacteria, viruses, fruit flies, and mice—their work was revealing an astonishing and deeply satisfying unity among life-forms in such matters as the genetic code, gene regulation, and biochemical pathways. They worked as teams in highly organized laboratories generously funded by grants. With money came space, power, prestige, and arrogance. The work of ecologists was dismissed as trivial, concerned only with the particulars of what was

disparagingly branded natural history. Paleontologists fared little better at the hands of their geological colleagues.
Geophysics and geochemistry emphasized principles of great
generality, whereas the old-fashioned fossil hunters, who in
any case were intellectually ill-equipped to handle quantitative
science, were content to play with shells and old bones. The
message was clear. If those who insisted on studying whole organisms wanted respect, they must transcend the specifics and
seek unifying principles with implications beyond the bounds
of their specialties.

There is much to despise about such holier-than-thou attitudes and about the destructively competitive system that produces them; yet they contain a grain of truth. Many traditional
biologists were indeed parochial in their outlook and cared little for the big picture. In response to the descriptive, detail-
oriented tradition, by the early 1970s many practitioners of
the beleaguered fields had jettisoned the data of natural history
in favor of mathematically sophisticated abstractions, which
sometimes bore little resemblance to the real world they were
meant to describe. Robert MacArthur had even maintained
that ecologists should dispense with the names of species. Such
details, he argued, obscure the generalities that theory could
reveal.

I wanted to understand the real world, with all its diversity
and complexity. Natural history provided the data that must inform and constrain the model-building. If we knew how organisms work, how they interact with one another and with their
inanimate surroundings, and how they are distributed in space
and time, we too could discern patterns and conceive ideas that
even our most ardent critics could appreciate. I also knew that
success in research was contingent on deep passions, which
must well up from a genuine and insatiable curiosity. Without
that passion, research would devolve into an intolerable

burden, a treadmill of writing and planning and executing difficult and often tedious work, always with the fear that one might run out of ideas or be spectacularly wrong. The only way out would then be to enter to one-way street of academic administration, in which one is ratcheted along from one bureaucratic post to the next in the trophic web until one is overwhelmed and permanently infected with meetings and paperwork. I wanted none of this. An incurable impatience with meetings and an intense dislike of bureaucracy, together with the conviction that administrators seldom benefit those whom they administer, ruled out this academic pathway for me. I believed I had the passion, the burning curiosity, the confidence, and the tenacity to make scholarly research the mainstay of my life at the university.

To do research, I had to find money. Readers, field assistants, and extensive travel would otherwise have to be paid for out of my modest salary. So would the charges that many scientific journals levy for publishing papers, as well as all the other large and small expenses. Reprints of published articles cost money, yet without them one cannot exchange information easily or acquaint others of one's work. I therefore had a powerful incentive to draft a grant proposal to the National Science Foundation even before the start of my formal duties at Maryland.

In today's competitive academic environment, making a sound choice of research direction is critically important. At most universities, a new faculty member has five to six years to become established before the fateful decision on granting tenure arrives. This is the interval during which one's work habits and pattern become essentially fixed. The easiest choice, and almost always the wrong one, is to follow along the path blazed toward the Ph.D. degree. There is a strong tendency to tie up loose ends, to bolster earlier conclusions with still more evidence, and to explore side issues. For a while, such a course might prove fruitful, but increasingly the work yields diminishing returns. By the time one is prepared to slay a new dragon on unfamiliar ground, so much time and effort are required to become knowledgeable enough to write a credible proposal or a meaningful paper that the tenure clock has sounded its farewell chimes.

In my case, there were plenty of loose ends left from the dissertation, but continuation along the Ph.D. path was not an attractive option. I could perform experiments, but the work would have limited appeal to my colleagues. In fact, even while I hammered away at the dissertation, I was becoming seduced by other projects and questions that opened up much larger issues. The question was not whether I would change my research emphasis, but how. Would I work on leaf shape in plants? Yes, but probably not as the primary focus. In the end, two projects for which I had been collecting data and ideas off and on for years looked especially promising.

The first of these belonged to a largely European tradition, that of describing biological form in abstract mathematical terms. In their spiral shapes, the shells of snails, clams, and their relatives can be thought of as variations on a very simple theme, that of an expanding coiled tube, or cone, wrapped around an axis. The point of the cone forms the shell's apex and is the oldest part of the shell. The wide end of the cone is the shell opening, or aperture, the part of the shell that is most recently formed by the molluscan builder. As the mollusc grows, it adds skeletal material to the rim of the opening, which extends in a spiral as it expands. Because the shell grows by adding material at one end only, rather than at two or more sites, it can be likened to a diary of the builder's life. The shell faithfully records the progressions of the seasons, the availability of food, and the onset of maturity, as well as the more unusual events that punctuate a molluscan life, such as a failed predation attempt. To make sense of the bewildering variety of shell shapes and to provide a common language for the description of shells, it is convenient to establish a geometrical frame of reference, a theme with variations.

At Princeton, Egbert Leigh had introduced me to D'Arcy Thompson's enduring two-volume treatise, *On Growth and Form*. In this extraordinary work, the exceptionally literate author explored with deep insight and frequent reference to untranslated passages from the Greek philosophers the mathematical and physical laws governing organic form. He drew attention to the logarithmic spiral, a mathematical curve in which successive coils of the spiral expand exponentially. The spiral shape of the shell, Thompson pointed out, closely approached the form of the logarithmic spiral. Most important, Thompson showed that the logarithmic spiral form arises whenever an expanding tube grows only at one end, as the shell of a mollusc does. There is, in other words, a logical connection between the observed spiral architecture of shells and the properties of the logarithmic spiral.

Alfred Fischer, meanwhile, urged me to read a new paper in the *Journal of Paleontology* by David M. Raup, who approached the mathematics of shell shape by means of computer simulation. Using only four quantities, or parameters, he was able to reproduce in the abstract a very large number of observed shell types, as well as many forms that do not occur in nature. With the parameters held constant as the shell grows, the coiled structure generated on the computer strictly conforms to logarithmic spiral growth. If the parameters are allowed to vary as the shell grows in size, subtle departures from logarithmic spiral growth become evident making the shells on the computer even more realistic.

Thompson and Raup had not only hit upon a realistic and elegant way of describing the known range of shell shapes encountered in nature, but had also identified geometries that do not, and never did, exist. Their methods made it possible to ask questions never before posed. What factors and circumstances set limits to biologically feasible or attainable forms? What makes some geometries forbidden in today's world or in the past? Why do some shell shapes never evolve? Are the limits imposed on shell shape today different from those in the geological past? What can be learned about an animal's physiological and biochemical machinery involved in the formation of shells from the underlying abstract mathematics?

I began to ponder these questions during my time at Yale. Having brought my collection from home to New Haven, I was able to consult it on a daily basis for insights into the geometry of real shells. Yale's Peabody Museum, moreover, possessed fabulous collections of fossils that further broadened my knowledge of the variations that had existed in earlier ages.

In March 1969, the removal of an inconsequential but painful cyst forced me to spend a few days in Yale's infirmary. With little else to do but catch up on sleep and think about shells, I turned over all kinds of snails, clams, and other shell types in my mind. Thompson's and Raup's descriptions captured much of the geometrical diversity available to molluscs and other shell-bearing animals, but not all of it. They accounted for variations in the shape of the shell's opening, the tightness with which the coils are wrapped around the shell's axis, and the rate at which the coiled tube expands as it spirals; but they did not allow for differences in the orientation of the shell's opening relative to the shell's axis. In theory, the axis around which the shell's coils are spirally wrapped can intersect the plane of the aperture at any angle. In my mind's eye (or hands, to be more precise), I inspected shells in which the axis lay almost perpendicular to the opening, as well as those in which the axis lay parallel to it. Quite suddenly, my mental groping caused observations to fall into place, making sense out of what only moments before had been hidden in murky confusion. Primitive snail shells were coiled with the axis steeply inclined to the shell's opening. This geometry, together with the fact that shells grow only by adding skeletal material at the aperture's rim, greatly limits the permissible variety of apertural shapes. The opening could be broadly oval to circular, but not narrow and long. More advanced shells usually have the axis nearly parallel to the opening, which can vary in shape from circular to narrowly elongate. The change in the orientation of the opening, in other words, permits a vastly greater range of shell shapes.

Upon release from the infirmary, I checked real shells against the abstractions of my imagination. All the shells but one followed the rules. *Cyclope neritea,* a small smooth snail from the Mediterranean Sea, seemed to be at odds with the pattern. Its axis rose steeply above the plane of the adult shell's aperture, which was long and narrow instead of elliptical as it should have been. The paradox was resolved when I realized that this forbidden architecture characterized only the adult shell. Like many other shells, that of *Cyclope neritea* ceases to grow when its builder reaches adulthood. A unique attribute of the species is that, as the shell approaches the adult condition, the angle that the axis makes with the opening changes. In early growth stages, the axis is almost parallel to the expanding aperture, which is long and narrow. At adulthood, however, the shell's axis tilts up steeply relative to the opening, which remains narrow. This geometry could not be achieved if the adult shell continued to grow by adding skeletal material to the opening's rim. The apparent exception therefore elegantly and convincingly proved the rule.

By the time of my job interviews, my first paper on this rather esoteric topic was on the verge of appearing in the *Journal of Zoology,* and a second was nearing completion. I suspect that having this second line of research firmly in hand greatly improved my chances of being interviewed.

Could the results from the work on snail shells be generalized? In my extensive reading on vertebrate and plant anatomy, I found reason to believe that the more recently evolved grand themes of organic architecture allowed for a greater range of variation than did the older, more constrained themes. I thought deeply about mounting a serious effort to investigate this hypothesis further, but in the end I chose to write up the preliminary findings and not to pursue the work. To describe shape in abstract mathematical terms is one thing, but to translate such a description into the molecular and organizational mechanisms that underlie the genesis of form in organisms is quite another. I had to acknowledge that I lacked the technical sophistication to handle the topic properly. What talents I possessed lay in work in the field, library, and museum. Much as I might have wished otherwise, I am neither a gifted experimentalist nor a competent mathematical modeler. My tastes lie in observing creatures in the wild, in comparative biology, and in the accumulation and evaluation of information from diverse sources. There is a point to improving one's weaknesses, but not to forfeiting potential advantages.

Even in the realm of observation-based field and museum work, some groups of organisms might have been far more suitable than molluscs for answering important questions; but it would have been foolhardy to throw away the love and knowledge I had of these animals in favor of barnacles, singing insects, plants, or other creatures that held much promise. Besides, molluscs presented a number of striking advantages. A great deal was already known about them, both as living animals and as fossil remains, so that one could ask many questions that for other less well known or less easily fossilized groups were intractable. Just as important, molluscs enabled me to make observations without help from others. Most move slowly, and they are generally easy to handle and to collect. Their diversity of form is so great that running out of interesting things to do is inconceivable. Finally, as I was to discover in the coming years, mollusc-based research can lead quite naturally to broader considerations involving detailed examination of such other groups as crabs, barnacles, fish, and even land plants.

One of the great benefits of the dissertation years was the opportunity to visit shores in many parts of the world. Despite the fact that my study animals lived on the upper reaches of the shore, where life is rigorous for land and sea creatures alike, I could not resist the temptation to wander down to the zones where the marine influence grew progressively stronger. On the lower shore, one encounters a lush world of seaweeds, barnacles, sponges, and a bewildering variety of molluscs. Such diversity holds an irresistible attraction for me. Why are some shapes, such as cone shells with a short spire and long narrow aperture, so frequently evolved in groups of tropical snails and wholly absent in freshwater organisms? What are the geometrical limits of form among molluscs inhabiting rock surfaces exposed to wave surge, and how do those limits differ from those among sand dwellers, mud burrowers, or snails living on the sheltered undersides of rocks? In short, I am intrigued by biological extravagance. It is as important to ask where such richness is found as it is to ask where the diversity of shapes and the number of species is modest. Just as we can know good only when we know evil, questions about diversity and biological specialization can be answered fully only if we investigate circumstances where they prevail and those where they do not.

I had been well prepared to engage in comparative studies. At Princeton in the fall of 1966, the newly arrived Egbert Leigh delighted in showing me the shells he had collected in the Red Sea, Florida, and Panama. Rather than arranging them in taxonomic order—all the species arranged by formal taxonomic groups such as genera, families, and orders, as I and most other collectors would do—he laid them out as assemblages of species that lived together in the same habitat, irrespective of the formal groups to which they belonged, so that any difference or similarity among regions was readily discernible. He drew no particular conclusions from this approach, but I was intrigued by it. Besides, it reinforced MacArthur's lectures on leaves. Throughout the tropics, MacArthur pointed out, the leaves of rain-forest trees conform to a narrow range of shapes, whereas temperate hardwoods show more variation, often having thinner, strongly toothed or serrated leaves rather than smooth-margined oval ones. Such comparisons can reveal patterns that demand an explanation; they might elicit questions that would not come up on only one site.

The limited regional comparisons I had been able to make on snail assemblages during my travels were beginning to reveal unexpected patterns. Whereas MacArthur and his predecessors saw uniformity in the leaves of canopy rain-forest trees, no matter which family the leaves belonged to or in which of the three great tropical biomes the trees grew, I was seeing subtle architectural differences among regions. Shallow-water snails from the western Pacific and Indian oceans were, on the whole, more strongly sculptured and had markedly smaller and more constricted apertures than their Caribbean counterparts. Shells from the tropical eastern Pacific fell somewhere in the middle. Rain-forest leaves converged in form, so the argument went, because functional demands and the evolutionary selection that molded leaf shape were similar throughout the canopies of the wet tropics. If that argument was valid, did it mean that selection and function in my snails differed from place to place?

With this question firmly in mind, I decided shortly after my dissertation defense to submit a research proposal on the architecture of tropical shallow-water snail assemblages to the National Science Foundation (NSF). I already had preliminary data for three of the four major tropical regions, the Indo-Pacific, eastern Pacific, and Caribbean. I had no firsthand knowledge of the fourth region, West Africa, although I had read extensively about its impoverished marine fauna and flora. Compared to the others, its fauna is notably small, but at the same time its waters are highly productive, so that the rocky shores, sandy beaches, and mud flats teem with a wealth of marine life. I also wanted to sample the other three regions more thoroughly. The Indian Ocean in particular would be important, because it represented a large part of the Indo-Pacific region that I had not yet visited.

The proposal was on its way to the foundation by June, but even if I were successful in securing funds, the money would be months away. Meanwhile, there was the summer to worry about. I had neither means of support nor any prospects of spending these months in the field. Out of the blue, however, John Corliss came to the rescue. He called one day in April with the news that my starting date would be moved up from September 1 to July 1. Not only would this enable me to begin drawing a salary earlier, but he would also try to persuade the university's graduate research board to help fund any research I might like to do over the summer. This was good luck indeed. Now, even if my request for funding from NSF were turned down, I would have something to show for the summer, as well as data to work up during my early months at Maryland. I began preparations at once for a six-week trip to West Africa.

When I asked Kenneth G. Rose if he would be interested in accompanying me, he jumped at the chance. A senior at Yale, Ken was a shell enthusiast who also displayed a keen interest in vertebrate paleontology. I had met him at Silliman College, one of Yale's undergraduate dining colleges, where Ken resided and where I regularly took my evening meal. In 1969, I had been chosen to take part in a program in which graduate students were encouraged to dine in the colleges with the undergraduates. Each recipient received one hundred dinners free of charge. To a person as short of money as I was, this benefit was substantial, made all the more appealing by the uniformly high quality and large quantity of food. Silliman was the closest college to the science hall, and a pleasant walk from geology along Hillhouse Avenue and across Trumbull and Grove streets. Jim Porter had also lived in Silliman, and it may have been through him that I met Ken.

By prior arrangement, Ken and I would visit marine laboratories and universities in Senegal, Sierra Leone, Ivory Coast, and Ghana. Everywhere we went we encountered extraordinarily helpful people who would accompany us to field sites, help us obtain permission from the local chief to look for snails on shores under their control, and arrange for our meals and lodging. Jacques Laborel, an expert on the coral reefs of Brazil and West Africa, graciously allowed us to stay in his house in Abidjan while he was away in France. In Ghana, Malcolm Edmunds and his wife Susan, both of whom had worked with the local molluscan fauna, guided us to promising collecting spots. Equally helpful was Walter Pople, and eccentric and somewhat racially unreformed South African with whom we had spirited arguments about his early sociobiological views.

The climax of the trip, however, came in Sierra Leone. A day after Ken and I arrived in Freetown, a hot and dusty city with the deepest roadside gutters either of us had ever seen, we checked into the Brookfields Hotel. This pleasant facility, located near the Presidential Lodge a short distance out of the main city, was one of two hotels where foreigners regularly stayed. It was visited every day by reporters from Radio Sierra Leone, who were eager to interview newcomers for a daily show called "Latest Arrivals." During the three-minute interview conducted with me, the well-spoken young reporter learned about my interest in the West African marine fauna and its relationships to other tropical faunas, but he may never have known I was blind, for he made no inquiries about my condition and did not remark on it. This was in any case an auspicious beginning to an interesting stay in a wonderfully diverse and exotic country. Not only could one listen to interviews with foreigners, but there were also news broadcasts from the BBC, Voice of America, and Radio Moscow, to say nothing of news in such languages as Mende, Temne, Limba, and Krio. Later that day, we ventured out to Whiteman's Bay for our first look at the local snails.

The following morning dawned cloudy. With the low tide not until late in the day, we put off fieldwork until the afternoon, planning instead to spend part of the morning at Fourah Bay College, English-speaking West Africa's oldest institution of higher learning. After cleaning specimens collected the previous day, we returned to Freetown by taxi. About halfway back, the driver came to a stop. A police officer sauntered up to the open window, peered in, and asked politely, "How are you? How long have you been in Sierra Leone, and what are you doing here?" As we explained about shore snails, the officer grew suspicious. "Do you have papers of identification?" he asked with a certain urgency. No, we acknowledged, they were back at the Brookfields. "Then I must detain you for interrogation. Please come follow me. I am not arresting you; I only detain you for interrogation."

We followed the officer to a roadside bench at a hastily organized police checkpoint. Again he inquired about our identification. The papers are at Brookfields Hotel, we assured him. Where? Brookfields. Did this mean we really had no identification papers? No, it simply meant we weren't carrying them, because they were back at the hotel. Were we associated with a government agency? Yes, we said, we are visiting Fourah Bay College and the Institute of Marine Biology and Oceanography. Where? We repeated the information. What were we doing? Research on shells, we assured the man, who by now was more than a little frustrated with the answers he only dimly understood. I produced the shells we had just cleaned from the pockets of my shorts. As I did so, I remarked on the radio interview of the previous day. Moreover, I pointed out, carrying passports with wet shells was a bad idea. At last, and none too soon for us, the officer relented. "Very sorry for the detainment," he said apologetically, and sent us on our way in the taxi whose driver had rather miraculously and bravely waited for us.

Officials at the American embassy had a ready explanation. Earlier in the day, the country had been thrown into a state of emergency following a failed attempt on the life of Sierra Leone's president. There were rumors of rebels streaming in from the northern provinces and from neighboring Guinea, and a dusk-to-dawn curfew would be clamped on the city. We happened to be traveling south toward Freetown, and as potential foreign mercenaries we quite naturally fell under suspicion.

With the political situation so unstable, we decided to move our departure to the Ivory Coast up by a few days, but not before visiting a few more rocky shores and mangroves. Like most of West Africa's coast, that of Sierra Leone consists of a succession of wave-exposed and often very steep sandy beaches, interrupted here and there by sand-scoured rocky headlands and muddy mangrove-lined estuaries. Relatively few species occur in such habitats, and those that do, by tropical standards at least, have rather plain shells.

Even if the shore molluscs of West Africa were not eye-catching, Ken and I uncovered some surprising species. One of these was a small, smooth, spotted periwinkle known as *Nodilittorina meleagris.* This tiny snail, which is very common on surf-beaten midshore rocks in the West Indies, was un-known from West Africa until we found specimens near Tema, in eastern Ghana. At Takoradi and Dixcove in western Ghana we found several small species that Serge Gofas would name as new species ten years later.

Eventful and interesting as the six-week trip to West Africa was, I thought often of Edith and wrote postcards and typed letters to her on borrowed typewriters at every opportunity. I wanted to propose to her, and I hatched a plan that I hoped she would agree to. Upon my return, we would take a trip to Maine where, surrounded by north woods and the sea, I would spring the question.

In early August, a few weeks before my move to College Park, Edith and I spent an idyllic six days on the Maine coast near Boothbay Harbor. This was Maine at its best — quiet cool spruce woods with inviting cushions of soft moss, sunny patches with wild blueberries and raspberries ready for pick-ing, splendid cold rock-pools rich with kelps and mussels and snails and sea stars, irresistibly delicious fish and fresh pies — a perfect place to be in love. Even the mud and the hordes of vi-cious mosquitoes we encountered during Edith's first visit to a proper salt marsh did not spoil the mood. The uncertain ice of the winter had given way to firm ground. When Edith chose to postpone her reply to my proposal of marriage, my disappoint-ment was tempered with the knowledge that our relationship had reached a stage of strong mutual commitment and affection.

I began work at College Park in late August. Although my first days there were hectic—teaching, advising students, unpacking, resuming my work on several papers, and incorporating specimens from West Africa and Maine into my collection—worries about funding were never far from my mind. What, I wondered aloud to Edith as we sat on a bench on a glorious November day in Washington during one of her visits, would I do in the likely event that my grant proposal were rejected? The National Science Foundation would probably consider the work too offbeat, too much of a gamble, and insufficiently experimental in its design. I was proposing to look for patterns and not to test carefully framed hypotheses as was even then de rigueur in outlining a fundable research program.

My pessimistic state of mind did not improve the next morning when Edith's commitments called her back to her cell cultures and experiments in the lab in New Haven. Her leaving always created a void, but on that morning my sadness over her departure was compounded by the necessity of attending the dreaded monthly faculty meeting. I could look forward to at least two hours of discussions of committee reports and much else that could have been handled more expeditiously in a memorandum. Sensing how little was accomplished during the first meeting in September, I blithely skipped October's gathering; but Corliss firmly reminded me that it was each faculty member's duty to attend, making my participation in the November meeting inescapable. I gritted my teeth and attended.

After lunch, as I was putting the finishing touches on a lecture I was to deliver at two o'clock, the telephone rang. "This is Ed Kuenzler from NSF," the caller announced. My heart raced, my hands quivered, and my fingers moistened as they grasped the receiver. "I have good news," he continued. My proposal would be funded almost entirely—all, that is, except for the fieldwork I had outlined.

There was no time for analysis now. Did the agency balk at spending money abroad at a time when there was much concern about foreign outlays? No, Kuenzler replied. The panel that judged all the proposals submitted to the Program of Biological Oceanography had concluded that a blind person would be unable to carry out the fieldwork. I knew I had a challenge on my hands. My dissertation, I told Kuenzler, could not have been written without many months of fieldwork on tropical and temperate shores. I had negotiated the razor-sharp cliffs on the north coast of Jamaica, surf-beaten slippery shores in Chile, and the treacherous cavernous algal ridge in Guam. With a field assistant, I should have no problems in the field, I continued.

Kuenzler listened politely and intently as I hastily laid out my case. Soon he interrupted. "I think that, on the basis of what you tell me, I will overrule the recommendations of the panel in this matter," he declared. Once again, I had the great fortune of an initially skeptical official who was willing to listen. He was open-minded enough to hear arguments and evidence and secure enough to set preconceptions aside.

The most pressing concern now was to find a research assistant who, on a permanent basis, could help in all aspects of my work. The NSF grant of $21,000 provided the means; all I had to do was to find the right person. Great good luck was again about to play its unpredictable but essential role. Neal G. Smith from the Smithsonian Tropical Research Institute in Panama stopped by Doug Morse's lab one day in early 1972, while Morse and I were chatting over lunch as had become our habit. With him was a tall, self-possessed woman, whom Neal introduced as Bettina Dudley. She was looking for a job, and did Doug know anyone who needed a hardworking person with an undergraduate degree in English from Cornell and a master's degree in botany with John Terborgh at Maryland. Bettina loved the outdoors and had even spent some weeks in the rain forests of Peru with Terborgh's party. My ears perked up at once. As soon as Bettina spoke, I knew she was perfect for the job. She had a beautiful, clear, full voice, exceptional diction, and a rich vocabulary. Wit and laughter came easily, a good antidote perhaps to my full-blown and by now incurable taste for puns. I was very eager to have her work with me, and after the briefest hesitation, she agreed. She would begin in September.

Sixteen years of a warm, happy working relationship followed. Bettina proved to be an exceptionally good reader, an accomplished editor, and a capable and energetic field assistant, as well as a close family friend. She was also deeply interested in education, and during her years with me finished a Ph.D. degree based on the study of the Woods Hole Children's School of Science, where she taught during the summer for several years. We engaged in many long discussions about education, and I can say with no exaggeration that Bettina's influence on my educational views was profound. Her sense of what students need and how they learn was far better than mine, and she seemed to understand as few others do how to motivate and encourage students.

Teaching was, of course, an essential function of a professor. Establishing a research program early was crucial if I wished to join the ranks of the faculty in line for tenure, but scholarship alone was not enough. If I did not measure up as a teacher, an academic career was out of the question.

My first assignment at Maryland, John Corliss informed me, was to teach a course in zoogeography in the fall semester of 1971. This was welcome news. The field continued to be as exciting then as it was when I was introduced to it by MacArthur and Fischer five years earlier. With the publication of MacArthur and Wilson's *Theory of Island Biogeography* in 1967, the descriptive tradition of identifying and circumscribing biogeographical provinces was giving way to the realization that patterns of distribution arise from processes of speciation, extinction, and invasion. Moreover, there was renewed interest in geographical patterns of species diversity. Why are there so many species in the tropics as compared to the higher latitudes? To what extent are present-day ecological factors such as habitat area, productivity, and competition responsible, and how has history influenced the number of species? The new theory of plate tectonics was also sweeping away older notions of fixed continents in favor of a more dynamic view of the earth's crust and its biota. Continents and oceans are not static features, and neither are the communities of plants and animals that inhabit them. Barriers limiting the ranges of species come and go, so that plant and animal distributions change dramatically as the configuration of land masses and ocean basins changes along with climate and tectonics. Much of the field had been dominated by those who studied birds and other terrestrial forms of life. I hoped that my marine perspective would enrich the course as I set about writing lectures during my last few months at Yale.

Unlike my peers, I was as green as the plants I intended to feature in many of the lectures. Of course I would make mistakes, but the esoteric nature of the field would ensure that only a few people would suffer, or so I thought. The first day of classes rudely disabused me of this fantasy. As I entered the long narrow lecture hall, aptly named the Shoebox, nearly one hundred students poured in, ready to take in my carefully rehearsed first remarks.

More than half the students were older than I was. One woman, disgruntled over the midterm examination she had just failed, showed me just how much I had to learn about teaching. "Young man," she said, "I have been teaching at the junior college level longer than you have been alive, and you can't just fail me like that." Well, I could; but I learned to be far more tactful with unacceptably poor performance.

As the first semester drew to a close, the zoology department was ready to recommend that my position as instructor be converted to a tenure-track assistant professorship. With the university administration's approval, I joined the ranks of the regular faculty in February 1972. Grant in hand and with still another new course to teach, this time on function and form in organic design, I was now on an equal footing with the two other assistant professors who had joined the department that year.

Douglas E. Gill had arrived from the University of Michigan, brimming with enthusiasm for theoretical ecology and the power of experimentation. One of the most energetic and perceptive natural historians I have ever met, he was a man bursting with personality and talent. I could readily identify him even in the absence of his superbly inflected bass voice, for he had a distinctive quick gait punctuated by the clop of the soles of his shoes in rapid succession as his feet came to a sudden halt in front of my always open office door.

Among Doug's passions was music. In another life he might have sung the role of Wotan in one of those interminable, overly romantic Wagner operas that he and his wife Sandy loved so much, but in this world he sang with the Washington Performing Arts Society. From this sophisticated city role, Doug could metamorphose into a dedicated field biologist, taking frequent grueling trips to the mountains of Virginia. Here he soon discovered the merits of long-term monitoring of populations of newts, dragonflies, orchids, and other creatures that lived in or near a complex of small ponds. Snapshot sampling was imply inadequate to expose the pattern and underlying causes of great variations in population size. Using the unique spotting pattern of individual newts, Gill was able to track the movements of adults and to show that these small amphibians exist for years in the terrestrial eft stage. An important implication of this work for me was that, although plants and animals are generally suited to their environment and evolution has adapted them to it, such adaptation is unlikely to be terribly precise. Populations and environments vary so dramatically and often over such short intervals of time that adaptive precision can hardly be expected.

The other new recruit was Richard R. Strathmann. His knowledge of invertebrates, especially their larval forms, rivaled Willard Hartman's, and he applied this store of information to a bewildering array of questions that few other biologists had even considered, much less tried to answer. Strathmann loved ironies, and he reveled in finding inconsistencies between what animals actually do and what they are thought to do. He was an ardent believer in careful observation, and his work on feeding by larvae and on filter-feeding by adult invertebrates amply demonstrated the power of his approach.

Strathmann and I quickly fell into a routine of visiting the library to inspect the newly arrived periodicals for interesting papers. It was on one of these weekly sojourns that I learned about his wooden leg. Hearing a repeated chirp, I remarked on the unusual cricket. "No," Strathmann replied, "it's my leg." Early in his college years, he had fallen off a moving freight train, and one leg required amputation. The squeaky hinge of his replacement leg now became his trademark. His condition did not prevent Strathmann from diving or from negotiating steep rocky terrain, as Edith and I were to discover in 1976 when he took us to the outer coast of Vancouver Island. By that time, Strathmann had become the assistant director of the

Friday Harbor Laboratories, one of the most idyllically situated marine biological facilities in the temperate world. My admiration for Strathmann and his deeply original mind has not diminished since 1973, when he left Maryland to take up his Friday Harbor post.

One event was essential to complete my transformation from the carefree ways of a student to responsible adulthood. I was more resolved than ever to marry Edith. Undeterred by her uncertain reception of my first proposal in Maine in 1971, I was determined to try again during our grant-supported research trip to the Indian Ocean during the summer of 1972. Edith and I visited several sites in Kenya, Madagascar, and Israel to acquaint ourselves with the molluscs of the reefs, mangroves, sea-grass meadows, and rocky shores of the western rim of the vast Indo–West Pacific Province. It was on the island of Nosy-Be, in northwestern Madagascar, that I proposed to Edith the second time. On this occasion Edith unhesitatingly accepted. The only question now was when and how the ceremony would be conducted. There would be legal obstacles if we married in Madagascar or in Holland, where we stopped on the way back from Israel to the United States. We did, however, exchange rings during our stay in Holland, having bought them in Gouda on the seven hundredth anniversary of that city's charter.

On October 28, 1972, Edith and I were married in a moving ceremony in Yale's Harkness Chapel, presided over by Justice of the Peace Harriet Seligsen. The high ceiling of the chapel sculpted the counterpoint of the organ and flute music of Bach and Vivaldi into a stunning work of acoustical architecture as we took our vows.

It was not until the following spring, when she successfully defended her Ph.D. dissertation, that Edith was able to join me permanently in College Park. Our weekends could at last be spent in tranquillity instead of on interminable train rides abroad the Metro-Liner between New Haven and Washington.

Tenure, when it came, was a welcome anticlimax. Before I had begun to worry about my prospects, it was granted in 1974. More than ever I now felt the freedom to devote myself to the life of a scientific scholar.

Chapter 10

RISKS, RAYS, AND RAMBUTANS

All the world's creatures live and evolve in a context. They are not little worlds unto themselves, isolated from one another and from the forces of wind, water, and earth. Instead, they persist, resist, respond, and perpetuate themselves in an environment rife with challenges and opportunities. Just as historians cannot expect to understand the rise and fall of nations or the court intrigues of seventeenth-century France without probing the economic and political motives of the human participants, so biologists seeking to document and explain patterns of evolution must penetrate and observe the world from the organism's perspective. Much can be learned from books, but the knowledge thus gained is inevitably filtered through someone else's faculties. There simply is no substitute for making one's own observations in the wild.

But is it reasonable to extend this necessity to a blind man? Isn't blindness, a condition nearly synonymous with helplessness, emphatically incompatible with experiencing unsupervised nature? The dangers are all too obvious—venomous snakes, stinging ants, poison ivy, moray eels, crocodiles, biting crabs, stonefish, stingrays, crumbling cliffs, freak waves, slippery rocks, deep crevices, choppy seas, menacing thunderstorms—the list could go on. None of these risks should be underestimated. I have been exposed to every one of them.

When Edith and I searched for shell-crushing crabs in Guam during the summer of 1974, we frequented a tract of boulders on the seaward part of the reef flat in Pago Bay. On one hot afternoon, I tilted a huge slab, steadying it in a vertical position with my leg as I gingerly inspected its lower surface with my left hand. There was the usual rich assortment of sponges, worms, and small snails, but there was also something else. Resting motionless on the rock was a smooth creature that yielded slightly under the gentle pressure from my fingers. It was not long or soft enough to be a sea cucumber, so I had no need to anticipate the expulsion of immensely sticky threads from the animal's insides, no need to free my fingers from this underwater glue. No, this creature was broader, flat, unmistakably fishlike; but there were no scales, and it remained still. Stonefish—the thought of that fearsome sit-and-wait predator flashed through my mind as I gulped. Edith, busy with a slab a few paces away, interrupted her search to see what interesting animal I had discovered. With a gasp, she confirmed my suspicion. Quickly but gently, I returned the slab to its original position, leaving the inhabitants of that marine underworld to carry on as before.

We both knew the outcome could have been disastrous. Had I brought my hand down more vigorously on the fish, it would have been provoked to erect its dorsal spine and to inject a potent neurotoxin that at the very least causes excruciating pain.

Sometimes I was not so lucky. In February 1986, I took Bettina and my family down to Panama to study burrowing snails. Immense sand flats on the Pacific side support hundreds of species of snails, clams, sand dollars, crabs, and other marine life. The largest and richest of these flats was Playa Venado, located west of the Panama Canal under the flight path of planes taking off and landing at Howard Air Force Base. One day, Bettina and I walked far out at low tide over the featureless sand until the water was deep enough to lap at our knees. Species that normally live below the low water line could be found here, and the exceptionally low tide made the area accessible by foot. One is sorely tempted to go barefoot on such inviting soft sand, but on this and most other occasions I wore sneakers. The pools through which we waded on the way to the edge of the beach were inhabited by aggressive swimming crabs, and the sand surface was littered with sharp-edged shell fragments. As I groped in the sand for snails to take back to the laboratory at the Smithsonian Tropical Research Institute, a sudden sharp pain in my right foot jolted me out of complacent preoccupation. Had an especially large crab pinched my toe? Upon quick reflection, this possibility seemed unlikely. A crab bites quickly and then lets go, unless it throws off, or autotomizes, its claw when threatened, in which case the claw clamps ever more tightly while the crab safely swims away. The pain persisted. As I reached down to diagnose the problem, I discovered a small stingray, flapping as it sought in vain to free its barbed tail from my foot. Gently, I tugged on the ray, which quietly swam off, leaving part of its tail embedded in my throbbing, profusely bleeding foot. Rays often lie motionless just beneath the sand, and when stepped on or otherwise disturbed will defend themselves by stinging the offender with the agile tail. Shuffling one's feet will usually prevent an encounter by giving the ray

early warning, but on this occasion I had failed to take that
sensible precaution. Fortunately, the ray's venom loses its effectiveness when heated. A knowledgeable nurse at the hospital promptly bathed my foot in scalding water, and the pain
miraculously subsided.

On another occasion I fell victim to a moray eel. In
August 1993, Edith, our daughter Hermine, and I paid a
brief visit to the Polynesian island of Moorea. The island is
situated in a part of the tropical Pacific with which I had no
previous firsthand experience. I was eager to visit the wave-
exposed reef edge, always a good place to find molluscs, and
an especially favorable habitat for *Pollia undosa*. This
species, its shell covered with a thick, fuzzy, organic periostracum, was of special interest to me because of my work on
snails such as *Pollia* whose outer-lip edge is marked with a
small protrusion that may stabilize the animal while it clings
to rocks. As usual, I snaked my fingers under ledges and into
holes, places where snails are apt to shelter. On this occasion, a pair of jaws belonging to an aggressive moray eel was
waiting for me. The bite was sudden, brief, and powerful.
Although the affected finger required a few stitches and initially felt numb, the attack left no permanent injury.
Ironically, the moray shared its lair with a beautiful adult
Pollia. The moray of this story is not to poke fingers care-
lessly into Île Moorea.

Fortunately, such brushes with natural dangers have been
as infrequent and inconsequential for me as they have been for
most of my sighted colleagues. Like anyone else who does
fieldwork, I take sensible precautions. I am keenly aware of my
surroundings at all times, and I never go into the field alone. It
is surprisingly easy for me to negotiate even the most jumbled
boulder beach or craggy ironshore when I hold a sighted companion's elbow. Once I come to a site I wish to investigate, I

work effectively on my own. Often, I am on all fours, or kneeling, acquainting myself with the topography and never moving swiftly unless the way is free and clear. When venturing around cliffs or onto narrow surf-swept ledges, I am always careful to keep a secure hold and to reconnoiter the terrain well enough so that I have a fall-back position in case of a freak wave. I keep a sharp ear out for the sound of the surf and for audible clues of crevasses, deep pools, and other potential hazards. As I work, my companion often strays far away, knowing that guidance is needed only when we wish to move quickly or far.

Edith has been my field companion on most occasions, but in recent years our daughter Hermine has also begun to play an active part. At a remote beach in Otago in 1993, Hermine and I were making our way above the low water mark to a surf-swept point of rocks. As we neared the point, a sharp musky odor with a faintly fishy undertone told us of something new nearby. Suddenly, Hermine stopped. "Oh, there are seals," she exclaimed. Large male fur seals, their skins adorned with the scars of battle, were draped on almost every large rock. Besides basking in the sun, they were busily engaged in aggressive provocations, and Hermine watched in awe as we took care to keep a safe distance.

Occasionally, I have been forced to rely on complete strangers as field assistants. In March 1986, UNESCO convened a highly stimulating workshop of ten scientists in Fiji to discuss the causes and economic implications of large-scale ecological differences among the world's great reef ecosystems. Field excursions on the island of Dravuni, where the University of the South Pacific maintains a small research facility, would allow participants to inject real observations into the deliberations. Everyone in the group except me dove. The fear of losing my hearing acuity had always discouraged me from diving, an activity otherwise perfectly accessible to the blind. While the others spent the day offshore in the water diving on spectacular reefs, I inspected the island's equally interesting shallows. A local fourteen-year-old boy proved to be a first-rate field companion and informant.

Surprisingly, concerns about safety have rarely been expressed by those who have paid for or have had to give me permission to conduct fieldwork. Most of the directors of marine laboratories where I have worked scarcely gave the matter any thought. When in 1978 we were making preparations to join a fifty-four-day research expedition aboard the R.V. *Alpha Helix* that would take place the following summer, I worried that officials at the Scripps Institution of Oceanography, which at that time operated the ship, would balk at my request. The ship would carry about a dozen scientists from Darwin, Australia to Cebu in the Philippines by way of Papua New Guinea and eastern Indonesia. It was the chance of a lifetime to visit places that would otherwise be unreachable, and I was more than a little eager to go. The matter was settled almost before I had time to map out strategy. A man identifying himself as one of the ship's captains called one day to ask if I thought blindness would pose a problem. I assured him that I had years of tropical marine field experience and that I had recently spent a wonderful week with the *Stenella,* a small vessel operated by the Smithsonian Tropical Research Institute, in a remote part of western Panama. The caller was satisfied and hung up, and the issue never came up again.

During the expedition, the ship's exceptionally cautious captain invariably anchored far offshore to avoid grounding the vessel. The ship's skiffs therefore had to travel miles over the open sea to reach shore. If this man had any qualms about my being part of the scientific party, he never expressed them. Running aground was evidently of greater concern to him than any risk I might pose. His confidence was well-founded. While several of my scientific shipmates received minor injuries of one kind or another, I remained unscathed save for a nasty little infection that developed on my leg after a minor coral scrape.

The voyage, however, was anything but uneventful. Having been built for sailing in rivers and shallow coastal waters, the *Alpha Helix* was notoriously prone to rolling. Edith and several others succumbed to seasickness as we plied laboriously across the Gulf of Carpentaria and, later in the voyage, negotiated choppy waters near the Aru Islands. Far more seriously, however, some containers of the toxic preservative formalin with which several colleagues were preserving copepods and fish fell from the shelves in the wet lab and shattered on the floor, filling the room with unspeakable fumes. People sometimes complained that the shells I was trying to clean smelled bad, but the formalin spill outdid anything I could do.

It was not the forces of nature but the whims of man that posed the greatest dangers. Early in the voyage we went ashore on a particularly desolate stretch of shore at Duyfken Point on the western side of the York Peninsula of northern Queensland. It was late afternoon when we climbed back into the skiff for the long bumpy ride back to the ship. The tide, having fallen considerably since our outbound trip, had nearly uncovered coral heads that before were well submerged. The crewman who piloted the skiff was unperturbed, however. He gunned the motor as we sped along. Within minutes, the outboard's propeller glanced a blow against one of the coral heads. Most boat operators would immediately stop, bring up the motor, assess the situation, and then proceed very cautiously; but not our man. He continued on as if nothing had happened. The inevitable second coral colony reared its head in our path, but this time the motor flew out of the boat, accompanied on its final voyage by a strong of oaths from our daredevil skiffmaster. There we were, motorless and still very far from the ship, near a forbidding stretch of lonely beach, with the sun about to set. Fortunately, one member of each shore party carried a radio, which now became our life line. The ship would send the second skiff to tow our disabled boat back. The operation was complete by nightfall, but not before we had gained an appreciation for the difficulties that early explorers must have faced in the days before outboards and radios.

The prospect of spending the night on an Australian beach paled in comparison to our worries during an incident in Indonesia. Edith and I had gone ashore by skiff near Korida, a small village on the island of Biak. This remote place is situated off the north coast of Irian Jaya, the western part of New Guinea that once was a Dutch colony but is now under Indonesian control. As we headed into a mangrove swamp, we were soon joined by the curious onlookers who so often turned up whenever we landed. In hopes of finding a more secluded spot, we boarded the skiff after half an hour, pulled up the anchor, and proceeded on our way.

Suddenly, two men armed with ancient guns pulled alongside in another boat. With others soon joining them, a rope was tied to our skiff, and one of our interceptors, none of whom spoke either Dutch or English, climbed aboard to direct us to the village's pier. Evidently, we were suspected of being Taiwanese poachers. At the dock, a headman speaking a smattering of Dutch appeared. Because I was the only one in the party capable of communicating with anyone in the village, I was ordered to the small police station. As Edith and I walked along, hundreds of jeering people lined our route. In the little concrete building itself, children were chased out of the doorway and dozens of curiosity-seekers gawked in the windows as I tried to explain to the dignified and polite headman what we were up to in the mangrove swamp. The chief demanded to see the ship's papers, and arrangements were made over the skiff's two-way radio to have the documentation brought.

The radio man aboard the *Alpha Helix,* as great good luck would have it, was Henry Folkerts, a sixty-eight-year-old Dutchman. He and I had already become good friends, for he spoke flawless Dutch and was teaching me some elementary Bahasa Indonesia, which he had learned as a boy on Java. He now came ashore with the papers and, most important, with an unencumbered ability to communicate. No, he assured the chief, we were not Taiwanese poachers; the ship carried scientists, including several Indonesians, who were studying marine life. The chief was satisfied, but insisted that a representative from the village accompany each of the two skiff parties during our explorations in the afternoon. With this agreeable arrangement, the afternoon's work proceeded without further incident. As we dropped off our local guests at the pier that evening, a large sack of delicious rambutans, soft hairy fruits with a sweep pulp and large seeds, was waiting for us as a parting gift.

We had two more encounters with gun-toting soldiers on the voyage. The first occurred near the Sangihe Islands in northeastern Indonesia. After a visit to a spectacularly diverse volcanic shore bathed by some of the clearest water Edith has ever seen, the ship prepared to sail to Davao, our first port of call in the Philippines. Toward evening, however, a dozen or so heavily armed men boarded the ship, and threatened to arrest the lot of us and take the ship back to Menado, the shallow port in Sulawesi we had left the previous day. By this time, Henry had taught me enough Bahasa Indonesia to enable me to tell the soldiers the basics of my work with snails and crabs. Delicate negotiations and a little money diffused this dangerous situation, but as always in such circumstances the atmosphere hung heavy with uncertainty and with the fear that an inexperienced and trigger-happy soldier would cause real harm.

That danger was even greater during the final encounter, which occurred on the east coast of Mindanao in the Philippines. While Edith and I found a beautiful deserted limestone shore in Pujada Bay near the Moslem village of Mati, our shipmate Frank Barnwell was put ashore in a nearby mangrove swamp, where he was studying fiddler crabs. He was soon discovered by curious villagers who became suspicious and summoned two soldiers armed with machine guns. When the skiff came to take Barnwell back to the ship, the soldiers came along, demanding to see the ship's papers and to ascertain whether the field parties had obtained permission to work on the local shores. On its way to the *Alpha Helix,* the skiff struck a rock, lost its motor overboard, and had to be rescued by the second skiff. Unaware of this unfolding drama, Edith and I continued to work on shore, increasingly anxious about the skiff that was to have picked us up hours earlier. When the second boat finally returned us to the *Alpha Helix,* the armed men were still aboard and it was not until hours later that the situation was resolved.

In 1981, the unusual opportunity arose to visit the northern Marianas, a chain of mostly uninhabited volcanic islands strung between Saipan in the south and the Bonin Islands of Japan in the north. The volcano on one of the islands, Pagan, had erupted on May 15, and Lu Eldredge was anxious to see what effects the eruption had on the marine communities of Pagan's shores and reefs. Resourceful as ever, Lu arranged for a small party from Guam to sail to Pagan and several nearby islands aboard the *Si-Ti-Si,* a privately owned sailing yacht. Roy Kropp, who would soon become one of my Ph.D. students, came along, as did several other students who were working toward master's degrees at the University of Guam's Marine Laboratory. Edith and I happened to be in Guam studying populations of *Strombus gibberulus,* one of the common sand-dwelling snails of inshore reef flats on the island. I had long been interested in the northern Marianas because their marine fauna remained extremely poorly known. Because Edith was pregnant with Hermine, she was unable to come. Lu, Roy, and I flew to Saipan to joint the rest of the party.

There is something indescribably delicious about the sweet smell of lush vegetation carried on a fresh evening land breeze as one approaches an island by sea. Guguan is as undisturbed by humans as any Pacific island can be today, and its unspoiled richness could literally be heard as the yacht's skiff inched its way toward a beach of lava boulders. Thousands of terns screamed overhead, and on shore the island resounded with the songs of endemic sparrows and kingfishers. Huge coconut crabs, their claws so powerful that they can open coconuts, scuttled in the underbrush. Once, these creatures would have roamed Guam, but they have all but disappeared from that island. The ravages of World War II together with human predation and, more recently, the destructive depredations of an introduced snake, extinguished most of the native species and squeezed the endemic forest into a few inaccessible areas of limestone.

The *Si-Ti-Si* anchored in Pagan's Bandera Bay a day later. We set up a temporary camp near the abandoned house of Ben Aldan, one of the island residents, who along with the fifty other inhabitants had escaped just before the eruption two months earlier. A thick layer of crumbly volcanic ash covered much of the island, but living creatures were thriving everywhere. Fast-growing beach morning glory and tall grasses poked their foliage through the rubble, and vast numbers of houseflies descended on us as soon as we stepped ashore. The marine fauna, too, seemed none the worse for wear. Floating chunks of pumice had scoured some corals and worn away the algal cover on some shores, but large adults of many species carried on as they must have done before the eruption. (To its inhabitants, this beautiful island must indeed have been the pumiced land.)

Several of the shores of Pagan were accessible only by sea and had to be approached with great skill and just the right timing to avoid swamping the skiff or smashing its occupants to pieces on the rocks. Pialama was one such beach. Situated on the southeast coast of the island, its slippery basalt boulders were battered by surf. Bruce Best, a fearless Vietnam veteran and an infinitely level-headed sailor, piloted the skiff to shore. "Jump," he barked, the bow lurching within inches of the rocks. I jumped, landing with both feet on the cobbles just above the waves. With great skill, Bruce spun the skiff around, avoided the rocks with the outboard, and sped back out to sea. To go through such a maneuver once was harrowing enough but it had to be done a second time, for the shore party had to be picked up and brought back to the mother ship. It is a testament to Bruce's seamanship that he accomplished it without mishap. These are situations in which one must place complete trust in someone else and follow orders at once and without question.

The fauna of the islands raised some intriguing questions. Some species that were absent or very rare on the mainly limestone islands of Saipan, Guam, and the other southern Marianas lived in large numbers on the northern islands. Others, such as *Strombus gibberulus* and *Vasum turbinellus,* showed just the opposite pattern, as they were abundant on the reef flats of the southern Marianas and absent in the north. The Smithsonian's collections, however, contained specimens of these species from the northern Marianas, many of which had been collected by Hank Banner and others during an expedition shortly after the end of World War II. I was puzzled by the differences in faunal composition between the northern and southern Marianas, but even more by the discrepancies between our own admittedly limited findings and those of the earlier expedition. Had we simply overlooked the species in Pagan and the other islands? That seemed unlikely, for earlier visits by Lu, Roy, and others from Guam had similarly failed to turn up specimens of the species in question. Perhaps the northern islands had changed since the war. That, too, seemed improbable. Human disturbance had been minimal, and no natural disaster had occurred that did not also affect the southern islands.

I was still pondering this minor mystery when Edith and I paid a brief visit to Coconut Island during our stop in Hawaii on the way to Guam. Hank Banner, though ill, happened to be in his office on the island that day, devoted as ever to his lifelong study of snapping shrimps. Did he remember the expedition, and could he shed some light on the problem? I asked. Hesitating only briefly, Banner began to recollect the event. Having returned from the northern islands, the research party was camped in a tent on Saipan when a large military truck backed up, scattering and destroying many samples. "We tried to put the collection back together as best we could," Banner continued, "but we could easily have made some mistakes." Some samples taken at Saipan or Tinian were evidently mislabeled as coming from Sarigan and other islands farther north.

A few months later, cancer claimed Banner's life. With coauthors Lu Eldredge and Allison Kay, I published a list of the molluscs of the northern Marianas, together with an analysis of dispersal patterns among oceanic islands and a rendition of Banner's fascinating account. The paper appeared in *Micronesica* in 1984.

The expedition aboard the *Alpha Helix* enabled me to realize a boyhood dream. As a child at Bussum, I had listened to Meneer Bontekoe read us a story about the island of Halmahera.

It painted a vivid picture of coconut palms on a hot sunny shore, their fronds clattering in the soft breeze. The name Halmahera had an irresistibly romantic ring to it, and the image of palms along a mysterious tropical beach evoked wonder about how a place could be so different from my own surroundings. Later, as I learned about tropical shells, I occasionally daydreamed about picking up such objects on this unimaginable island. My first opportunity came on July 14, 1979, when my graduate student Mark Bertness and I came ashore at Dodinga Bay, on the west coast of Halmahera. The sandy silt of the shore held more than I could have dreamed of. Here was exquisite and unbridled diversity: seven kinds of nassariid snails; five olives; and assorted cone, auger, and turret shells. There was a large population of ark shells, many with drill holes in the valves. Some of the holes had penetrated to the inner surface of the valve, which meant that the clam had been eaten by the perpetrator; but almost half the holes were incomplete. The culprit was *Lataxiena blosvillei,* a bizarre mud-dwelling muricid snail.

One paper and part of another came out of my encounter with this rich community of molluscs. My experience at marine biological laboratories and aboard the *Alpha Helix* and other ships lulled me into the belief that my blindness would not be raised again as a concern by anyone in the science establishment. I was wrong. Early in 1987, I received a call from David O. Duggins, a marine biologist I had come to know and like well during my sabbatical at Friday Harbor in the spring of 1986. David and two of his collaborators, Charles A. Simenstad of the University of Washington and James A. Estes of the National Marine Fisheries Service in Santa Cruz, California, were organizing an expedition aboard the *Alpha Helix* to study the effects of sea otters on coastal ecosystems in the Aleutian Islands of Alaska. Would I, David asked, like to come along? Rich Palmer from the University of Alberta would come as well, and the two of us could work ashore while the others made observations and conducted experiments underwater offshore.

This was an opportunity too good to pass up. I had first met Palmer in 1976 in Panama, as I shall tell in the next chapter, and had taught a course with him on form and function in marine invertebrates at Friday Harbor in 1986. He was a superb naturalist and thinker, and we worked extremely well together. Besides, I had developed a deep interest in the fauna of the North Pacific.

This rich fauna contains many species with very close relatives in the cold North Atlantic. Today, a marine connection exists between the North Pacific and Arctic-Atlantic basins by way of the Bering Strait, a narrow shallow seaway separating Siberia and Alaska. Before about four million years ago, however, a land bridge prevented any exchange of marine species between the two great northern ocean basins. Once the Bering Strait opened, a large-scale biological invasion ensued, in which hundreds of species spread from the Pacific to the Arctic and Atlantic, but only a handful spread in the opposite direction. I had become interested in this one-way exchange of species and wanted a firsthand look at a fauna that contained species similar to or identical with the ones that had extended their ranges into the Atlantic. At the time the Bering Strait became a seaway, the world's climate was substantially warmer than it is today, so that the fauna and flora living around the Bering Strait then were very much like those occurring today in the Aleutian Islands arc, 13 to 14 degrees of latitude to the south.

There was, Duggins explained, one small hitch. The expedition would take place aboard the *Alpha Helix*, whose operation had been taken over by the University of Alaska. Officials at that institution insisted that either my employer, the University of Maryland, or the University of Washington where Duggins and Simenstad worked, would have to sign a document in which any financial burden resulting from my being on board would be borne by either of these institutions and not by Alaska. Having a blind person on board, the argument went, would pose risks to me and to others. After all, Aleutian waters are inherently dangerous, and any work on shore would require landing in open skiffs on beaches where conditions were rough and unpredictable.

When I called Dolly Deater at the University of Alaska's Institute of Marine Science in Seward, my worst suspicions were confirmed. None of the other eleven scientists on board were required to supply such financial absolutions. She richly embroidered on the safety theme. Ship-to-skiff and skiff-to-shore transfers, she argued, were more or less out of the question for a blind person. She pointed out that one of the skiffs was swamped the previous summer as it attempted to land on one the islands. Once thrown into the frigid waters, one had only a few minutes to live. Even if I could persuade Maryland or Washington to relieve Alaska of its extra financial obligation on my account, Deater continued, the final decision about my participation would have to come from the University of Alaska's chancellor.

The full irony of this situation became apparent only later, as I reviewed my earlier voyage on the *Alpha Helix* in preparation for further action. It turned out that Dolly Deater was one of several members of a small delegation from the University of Alaska who came aboard the *Alpha Helix* at Thursday Island, Australia. The university was assessing the feasibility and desirability of taking over the operations of the ship for the National Science Foundation from Scripps. The party stayed on for a few days as we steamed to Port Moresby. Under way, we stopped at Ravao Island, where Edith and I rode one of the skiffs to shore. The trip was utterly uneventful. Nobody gave the transitions from ship to skiff, skiff to shore, shore to skiff, and skiff to ship a second thought. Why, then, was such a fuss being made about my participation on an expedition to the Aleutians?

To be sure, there are risks associated with any fieldwork. I was as anxious as anyone not to be tossed overboard into the unforgiving waters of the North Pacific, and I knew that medical help would be very long in coming in the event that I should slip on kelp or scrape myself on barnacles. There were no crocodiles, stonefish, stingrays, sea snakes, or cone shells to worry about, however, nor any venomous snakes or malaria on land. Besides, all the dangers that faced me would be faced by the others on the expedition. It seemed quite unreasonable to hold me or my employer to a standard of financial responsibility more stringent than that being applied to others. After all, the available evidence showed that I was capable of fieldwork under very difficult conditions and that I had not suffered more injury or posed more risk than anyone else.

There was no time to lose. Contacts with Marc Maurer, the energetic and extremely capable president of the National Federation of the Blind (NFB), confirmed that differential risk assumption with respect to blind persons was illegal under terms of Maryland's White Cane law. A similar law made the practice illegal in Alaska as well. In a long letter to various officials in Alaska, I set out my arguments in detail. I recounted the three previous shipboard expeditions, including the one to Pagan, and carefully laid bare the lack of evidence for my being a hazard.

Much to my relief, the officials at the University of Alaska gave in. We reached a compromise, whereby I was to supply Alaska with a letter from the dean certifying my competence in the field. The dean, of course, had never observed me in the field and was therefore hardly in a position to comment meaningfully, but in the arcane world of overcautious and paranoid administrators such minor details are of no consequence. The important victory was that my risk would be treated no differently from anyone else's, and I was free to join the expedition. I published a detailed account of this episode, including all the correspondence, in the NFB's magazine, the *Braille Monitor*, in 1988.

The expedition itself fulfilled all my hopes and expectations. After a long slow flight from Kodiak Island aboard a giant C-130 transport plane operated by the Coast Guard, we met the ship at Attu, the westernmost of the Aleutian Islands. It was hardly an auspicious beginning. A cold rain fell as we climbed aboard the skiff on the beach at Casco Cove, and a cold wind blew off the Bering Sea. The oddly familiar drone of the ship's engine drew nearer as the skiff tossed in the chop of the bay. I hoped the angry sea would not confirm the worst fears and doubts of the administrators in Seward. It did not. The skiff pulled alongside, and I scaled the short ladder to the heaving deck.

We spent the next day reconnoitering the lonely shores of Attu. At all times in the field, Palmer and I wore bright orange survival suits, heavy cumbersome overalls that would allow us to float and to stay warm for a while if we fell into the water. On the shore, I donned mittens with open fingertips, so that I could feel without numbing the whole hand. The water was so cold that it would immobilize my hand within minutes without such protection. One of our destinations that day was Murder Point, so named because of a Russian-led Indian massacre in the 1740s. "Et tu, Brute?" I muttered to myself, marveling at the contrast between this bleak place and Caesar's Rome.

Over the next two weeks, the ship sailed slowly eastward along the chain of islands that stretches for more than a thousand miles across the North Pacific. Palmer and I put ashore at eleven of the islands. Most are now uninhabited by people, and such places as Seguam and Chuginadak have rarely been visited by naturalists. On shore, the deep silence is punctuated by the odd sounds of birds: ptarmigan croaking in the dense grass, a bald eagle squealing overhead, or the tentative brief song of a snow bunting. There are no trees through which the wind can whisper, only a thick carpet of grass and cow parsnip. The ground is a thick spongy mat, immensely springy under the feet, where very little soil accumulates as generations of vegetation slowly decompose.

With so many landfalls, the chance of a mishap was considerable, but I was determined once and for all to erase all the doubts that skeptical administrators might still harbor. By my count, I made fifty-seven transfers between ship, skiff, and shore. On only one of these was there even the smallest incident. While climbing aboard a skiff at a beach on Amchitka, my right foot slipped on a kelp-covered rock, with the result that the hip boot protecting my leg partially filled with water. Rich Palmer suffered at least three such experiences. They were annoying, of course, but hardly life-threatening.

The Aleutian episode aptly illustrates the importance of assuming personal responsibility when one knowingly takes risks. In the long and mostly dreary history of the blind, protection against the outside world has been one of the grand themes. The blind are to be sheltered; they are wards and clients, to be kept away from stairways lest they should fall, from rosebushes lest they prick themselves, and from skiffs lest they drown. Safety is the most commonly cited reason for denying the blind access to amusement parks, skating rinks, cruise ships, escalators, laboratories, and jobs. Better to sit home and do nothing than to go out in the world, take a job, and earn one's way in society.

We cannot hope to know nature if we do not let ourselves be part of it. By throwing ourselves on the mercy of natural forces, we inevitably take chances, and we must accept responsibility for doing so. A risk-free world is a very dull world, one from which we are apt to learn little of consequence. We must have the right to try, even if trying sometimes results in failure.

Indeed, had I stayed home and not met the challenge of the Aleutian cruise, my understanding of the North Pacific fauna would have been seriously compromised. Before the voyage, I had thoroughly acquainted myself with the literature about the marine life of Alaska. Having spent several months in the Puget Sound region, where many of the species I would be seeing in the Aleutians also occur, I formed the preconception that the conditions of life in Aleutian marine communities would be similar to those in Washington and British Columbia. The first day at Attu showed how wrong I was. Most strikingly, the shells of Aleutian molluscs are thin, in great contrast to the often massive shells farther south. Moreover, the large crabs and sea stars that are so characteristic of the Pacific Northwest are notably scarce in the Aleutians. Whereas the shells of shore snails from Washington are often deeply pitted in their older portions, where boring organisms have excavated tunnels and cavities, the shells of Aleutian snails are remarkably unsullied by such damage.

These contrasts might not in themselves be meaningful or important, but there is no hint of them in the published scientific literature. As it happens, the thin shell walls of the molluscs and the scarcity of shell-crushing crabs reflect a generally minor role of shell breakage in the lives of Aleutian molluscs. This condition may in fact have been important in the ecology of the species that first invaded the Arctic Ocean through the Bering Strait from the North Pacific four million years ago. In the larger scheme of things, this sliver of insight may matter little, but it would have remained unrevealed to me had the well-meaning souls who took administrative responsibility for my well-being prevailed.

Chapter 11

CLAWS

I stand in ankle-deep water on a sloping rocky pavement. The sea slaps lazily against the skiff in which Edith and I have landed, anchored a few feet away. Below me, thickets of delicately branched corals, their skeletons peculiarly scabrous yet intricately sculptured, form a stony undersea meadow. Far above my head, on the densely wooded island, a concert of soprano cicadas is in progress, each insect singing fast little arpeggios like a miniature fire-engine siren in the blazing sun. The double tones of a bellbird, hauntingly pure and distant, echo among the trees and rock walls. I face a cave, hollowed out for a distance of twenty feet or more into the limestone foundation of the island by thousands of generations of rock-grinding bivalves and chitons. The tide reaches in, nourishing the thin veneer of mossy growths on the wet walls. On this mushroom-shaped island in the great lagoon of Palau in 1975, the luxuriant rain forest above touches a magnificent coral reef below, and I am fortunate enough to be in their embrace.

Edith and I have come to study death in this paradise. The unspoiled splendor and outward peace belie a world of danger for the creatures that live here. How could it be otherwise? Survival for animals means eating, and eating means that some will perish or starve while others live. The bright hues of the damselfish that I cannot see dancing among the corals may delight the eye, but they also bear witness to a struggle for mates and for the right to graze. Even the corals, those seemingly inert builders of the underwater stone forest, keep one another at bay by digesting or stinging their neighbors. I wanted to know how snails on this quarrel reef die, what agencies are responsible, and which attributes of the mollusc provide protection against potential causes of death.

That predators were instrumental agents of evolutionary change in snails seemed all too obvious now. The narrow openings could keep predators out or at least make entry more time-consuming and therefore more risky for the predator. The thick shell wall, sturdy spines and turbercles, tight shell coiling, and the ability of many snails to withdraw the soft parts deep within the shell were other features that complicated and slowed handling by potential enemies. The shell had become an effective deterrent to predators, to whom a wide variety of techniques—crushing, peeling, drilling, envenomation, extraction through the aperture, and swallowing the victim whole— were available. But it was not always so obvious, nor could I have predicted where the arcane study of snail-shell defenses might lead.

The first hint came in the form of a casual observation five years earlier. Shortly after arriving in Guam in 1970, Lu Eldredge took me to Togcha Bay, just as he had two years before. As we splashed through one of the hot tide pools in the July afternoon sun, Lu reached down, picked up a shell, and handed it to me. It was a pristinely glossy specimen of the money cowrie *(Cypraea moneta),* perfect in every way except that its domed dorsal surface had been broken away. I remarked that such crimes against nature seemed all too commonplace and expressed surprise that waves would be strong enough on this part of the reef flat to break such a sturdy shell. Lu offered a different explanation. He had observed crabs breaking shells in just this way in his aquarium. I kept the cowrie and noted the observation, but gave the find little further thought.

The broader significance of Lu's shell did not sink in until August 1972. On our way back from three months of rigorous fieldwork on the Indian Ocean coasts of Kenya and Madagascar and in the northern Red Sea, Edith and I stopped off to see my brother Arie and his wife Hanny. They were still living in their roomy flat in Zeist, a prosperous town in the center of Holland, east of Utrecht. With time to reflect on the summer's experiences, I finally saw the connection between shell shape and predation that should have been apparent years earlier. The molluscs of East Africa, like those in the western Pacific, were very strongly sculptured and generally had small openings. In Madagascar, I had found many broken shells, which I tossed away in disgust, but also some spectacularly damaged ones that the snail occupant had been able to repair. Most of the differences that I had perceived between the heavily armored shells of the Indo-Pacific and their less fortified Caribbean and West African counterparts might indicate a difference in the evolutionary role of predators. Perhaps shell-crushing predators were stronger in the Pacific and Indian oceans than elsewhere in the shallow-water tropics. Perhaps predation had influenced molluscs and other invertebrates for a much longer time there, with the result that victim species had been driven to ever more outrageous architectural specialization. Was this interoceanic pattern part of a much bigger picture? The intriguing papers of Gerald J.

Bakus, whom I had met briefly at Discovery Bay in 1970, came to mind. He had suggested that grazing by fishes on seaweeds was much more intense in the tropics than at higher latitudes and that within the tropics such grazing was especially vigorous in the western Pacific. It now seemed reasonable to propose the hypothesis that regional variations in shell architecture, which my first grant from the National Science Foundation had enabled me to document, reveal profound differences in the intensity and perhaps the history of predation.

By the end of the following summer, during which we sampled molluscs from the Atlantic and Pacific coasts of Costa Rica, I felt ready to commit this hypothesis and the supporting documentation of regional variations in shell form to paper. The result was published by *Evolution* in 1974.

Now came the much more difficult task of testing aspects of this hypothesis. Given that almost nothing was known about the enemies of tropical molluscs, the first problem was to identify the kinds of predators that were capable of breaking and entering shells. There were some clues in the published literature. In Ireland, J. A. Kitching and his colleagues had demonstrated that the common shore crab, *Carcinus maenas,* was capable of breaching the shell defenses of the dogwhelk, *Nucella lapillus.* More interesting, the thick-shelled, small-apertured form of the dogwhelk was better able to resist the onslaughts of crabs, with which it coexisted, than was the thin-shelled, large-apertured form that lived in wave-exposed habitats rarely frequently by the shore crab. Lu had also mentioned crabs as the potential culprits, but which of the hundreds of species in Guam were responsible remained unclear.

The obvious place to turn for guidance was the world's largest collection of crustaceans, located conveniently in Washington's National Museum of Natural History. Promising crabs would have large, thick-walled claws in which the prey could be squeezed or cut between large, molarlike protuberances. It was not long before we found some very formidable crabs. "Look at this," Edith exclaimed as she pulled open the lid of a tightly sealed jar and gingerly lifted out a huge crab, careful to drip the adhering alcohol back into the receptacle. The creature, some 12 centimeters across the back, was built like a tank. On the front side of its smooth domed shell, the animal carried two enormous claws, the right considerably larger than the left, the stubby fingers set with a row of broad low crushing teeth. This was *Carpilius maculatus,* the seven-eleven crab, so named because of the pattern of spots on its back. Specimens in the collection came from all over the tropical Pacific. Evidently the species was common on the reefs of many islands, including those of Guam. *Carpilius* was not alone. There were also massively clawed species of *Eriphia* and *Ozius,* although none was quite as clearly representative of the morphology I expect of a shell-crusher.

Guam was the ideal place in which to study these crabs and their snail prey. Since my last visit in 1970, the College of Guam had transformed itself into a university and added a marine laboratory on the windward side of the island above Pago Bay. The lab was situated near a particularly rich reef flat and had a functional system of running seawater, a few boats, excellent technicians, and a small but enthusiastic faculty with eager graduate students enrolled in a new master's degree program.

With a generous grant from the National Geographic Society, Edith and I set out for Guam in May 1974. A great deal was riding on the summer's work. If we could not find the right crabs or other predators with instruments capable of crushing shells, the research program that seemed so promising would wither. For Edith, the trip represented a golden opportunity to get back to her first love in science, the study of animal behavior. Ever since her undergraduate days at Brandeis University in Waltham, Massachusetts, she had been powerfully attracted to observing how animals behave and how the actions of an individual animal were related to the individual's habitat and form. Following the successful completion of her doctorate at Yale in 1973, Edith decided to give up molecular biology and its cutthroat competitive culture in favor of a less stressful, more flexible life of work in environmental science and policy. Her part-time position at the Environmental Defense Fund in Washington enabled her to devote summers to work in the natural environment she was committed to preserve.

It is obvious from inspecting a globe that the Pacific Ocean is a huge body of water, but there is nothing like travel to give real meaning to that fact. Almost six thousand miles of ocean separate Guam from California, and even Honolulu lies some eight hours of nonstop overwater jet flight and thirty-seven hundred miles from Guam. I can only marvel at the navigational achievements of Magellan, who in 1521 was the first European to set foot on Guam. Had he come from the west by way of the Philippines or the Malay Archipelago, he might only have had to sail a few hundred miles from the nearest land, but he came the long way, from the east, without a single landfall, some six thousand miles from the southern tip of South America.

The marine laboratory at Guam was just a one-story building in 1974. Built to withstand Guam's savage typhoons and cataclysmic earthquakes, it rose above the cliffs as a fortress of reinforced concrete with a heavy flat roof, a triumph of design completely dependent on the availability of cheap electricity. The architects of Guam's new buildings ignored the lessons taught by the snails of the upper tropical seashore. Where high-shore periwinkles and limpets would have a tall, peaked shell—with a sloping, peaked roof—to minimize the uptake of the sun's heat and to maximize its dissipation, the horizontal roof of the lab and of most of the other postwar edifices on the island soaked up the sun's heat efficiently and without mercy. Air-conditioning, so the theory went, would compensate; and so it did when the current was flowing, but there was many a day when the power failed and all the lab doors had to be flung open to alleviate the oppressive heat inside by allowing nature's cooling trade winds to blow through the offices and workrooms.

We rented a pleasant house on Dean Circle, some ten minutes' walk up an immensely steep, wooded hill from the lab. As we trudged up the hill, we first passed the Tenorio farm, owned by a respected local family with a reputation for putting spells on people. Then came the steepest part, where the road turned sharply to the left, through a dense thicket of tangantangan, a tropical American shrub that had been introduced by military authorities after World War II to combat the land erosion wrought by the war's devastations. The shrub grows like a weed and bears buds, flowers, and small flat pods simultaneously. It now covers vast tracts of the island and may be partly responsible for the high nitrate content of Guam's soil.

The Dean Circle house's usual occupants, on leave for the summer, were artist Joe Kagle and his family. The house was filled with a stunning collection of Palauan story-boards, intricately carved planks of *ifil* or Philippine mahogany depicting various legends from Palau. The carvers who created them had been encouraged by a Japanese anthropologist to transfer their extraordinary talents from carving the walls of *bai's*, or traditional men's houses, to long boards that could be sold to visitors. We were so impressed with this Palauan art that Edith and I acquired our own collection of story-boards, several of which we commissioned from master carvers in Palau and Guam.

The crab work got off to a very promising start. With the help of Dan Wooster and Rick Dickinson, two of Lu Eldredge's energetic master's degree students, we soon acquired several adult *Carpilius maculatus,* as well as some red-eyed crabs that we identified as *Eriphia sebana.* The red-eyes were smaller and more active crabs than *Carpilius,* but their master claws were nonetheless impressively long and powerful. Kept singly in aquaria with rocks for shelter and a range of sizes and shapes of snails for prey, the crabs quickly acclimated and set about crushing shells. Loud reports issued from the aquaria as the shells of victims gave way explosively under the pressure between the fingers of the crabs' absurdly oversized claws. For each potential prey snail, we kept careful records of its dimensions and its ultimate fate. Sometimes a crab could crush its prey outright, with only a pile of tiny unrecognizable fragments chronicling the attack. In other instances, however, the shell was so sturdy that an attack left much of it intact, enabling me to identify which features made the prey shell vulnerable and which ones provided protection. Both *Carpilius* and *Eriphia* often attacked prey shells first around the rim of the aperture. The margins of the shell opening were often so effectively reinforced, however, that this mode of attack by the crab failed. Ultimately, many crabs gave up the attempt, leaving behind a bruised and damaged shell with the unscathed soft parts of the snail safely and very deeply withdrawn inside.

Despite the remarkable defenses of many of the local snails, no snail species was wholly invulnerable to predation by crabs. Even *Drupa morum,* the most heavily fortified among the smaller shallow-water reef snails, was susceptible to crushing. One large female *Carpilius maculatus,* whom we affectionately called Railroad Tie for the heavy *ifil* wood that weighted down the lid on her aquarium, was able to crack the shell of a fully adult *Drupa morum.* This snail, which lives on surf-swept reef margins, combines nearly all of the features one might associate with effective shell armor. The shell, which is only about 35 millimeters (one and a half inches) long, has a hemispherical shape, with no protruding portions that a crab could snap off. The surface is adorned with thick, rounded knobs or tubercles, which make the 2- to 3-millimeter-thick shell wall even thicker in many places. The long aperture is only about 2 millimeters wide, thanks in part to large protrusions jutting part of the way across the opening from the rim. With the aperture so narrow and its rim so heavily fortified, the dome-shaped shell was uncrushable by most of our crabs, but Railroad Tie was powerful enough and large enough to hold the prey shell between the fingers of her huge right claw and to break through the shell wall opposite the aperture. Subsequent tests done in a force-measuring machine by John D. Currey at the University of York in England showed that a force of some 5 kilonewtons (about 1200 pounds) would be necessary to break a shell of *Drupa morum* of this size and wall thickness.

For all their power, the shell-crushing crabs we studied are docile creatures lacking the agility of the overtly more aggressive crabs that feed on other crustaceans or even on small fish. I had no fear of being pinched by *Carpilius* while I gathered up living snails or the remains of victims from the crabs' aquaria. Handling the crabs also posed little danger. With my fingers holding the claws firmly against the front of the crab's shell, I could lift and hold the crab securely while Edith cleaned its aquarium.

So little was known about the biology of even the most common reef species that unexpected findings came often. One day in early June, Edith and I waded into the moat area, a part of the reef flat of Pago Bay just shoreward of the reef crest. There large slabs and boulders of limestone are strewn by the thousands among deep, narrow, steep-walled channels and quiet pools. This was a favorite habitat of *Carpilius,* which lurked under large blocks during the day. Turning over a heavy slab that a long-forgotten typhoon had hurled onto the flat, Edith spotted a most unusual crab. It was an utterly motionless lump, looking for all the world like a piece of dead coral. Even its shelly carapace had a rough, stony texture, made all the more cryptic by growths of tiny encrusting animals. In the hope that Lu might know this crab, we took it back to the lab and put it up temporarily in a spare aquarium that served as a holding area for *Carpilius's* future prey.

Three empty shells, each with the spire neatly severed, lay on the bottom of the aquarium the following morning. Lu was not surprised. "Oh," he said in his matter-of-fact yet animated fashion, "That's *Daldorfia horrida.* We kept one in our aquarium for months, and it sometimes ate snails." In the folded-up position in which we found *Daldorfia,* the claws blend in so well with the rest of the animal that their size remained a well-hidden secret. In their extended pose, however, the claws were longer than the crab was wide, and the thick massive fingers bore molarlike protuberances ideal for crushing. *Daldorfia* proved to be a formidable predator, more apt to break shells in areas away from the aperture than were the other crabs.

The fact that crabs in captivity are able to crush snail shells does not mean that they are important either as agents of death in the wild or as enforcers of snail adaptation. I needed information about how molluscs died and, most important, about whether shells functioned successfully by withstanding the attacks of crushing predators.

Suddenly, all those ugly broken shells that I had previously dismissed as unacceptable specimens became mines of information. So did all the other shells—some in pristine condition, others in varying states of dilapidation—that served as homes

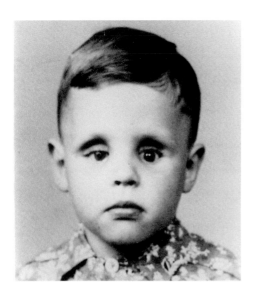

Photograph of the author taken in 1951, about 1 year after he became totally blind.

The author with his mother in front of the Institute for the Blind, Huizen.

Family portrait, 1959. The author's brother, Arie, is on the left; his father, Johannes, is on the right; and the author is standing in front of his mother, Aaltje.

Mrs. Ruth Saplow, third-grade teacher at Newton Elementary School.

Bettina Dudley.

Nerite snails, apertural *(left)* and dorsal views *(right)*. Above, *Nerita plicata,* a nearly spherical snail. This specimen is 25.1 mm in diameter and was collected from high-shore basalt rock on August 13, 1970, at Lalas Point, Guam. Below are views of a somewhat flatter species, *Nerita exuvia,* which live slightly lower on the rocky shore. This specimen is 34.4 mm long and was collected on July 18, 1979, from basalt rocks in the Sangihe Islands, Indonesia. (Photograph by Mary Graziose.)

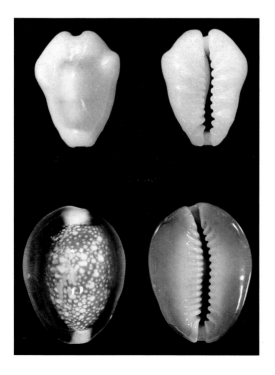

Above, *Cypraea moneta,* 17.6 mm long. This specimen of the money cowrie came from inshore rocks on a sheltered shore at Piti, Guam; collected on June 2, 1981. Below, *Cypraea caputserpentis,* 31.5 mm long. This snake's head cowrie was collected from a deep crevice in wave-swept rocks at the reef margin on Cocos Island, Guam, on July 30, 1974. (Photograph by Mary Graziose.)

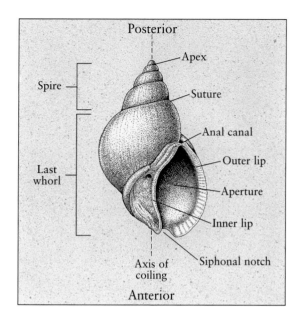

Typical gastropod shell, showing parts of the shell frequently discussed throughout the book. (Illustration by Roberto Osti, based on an illustration by Janice C. Fong.)

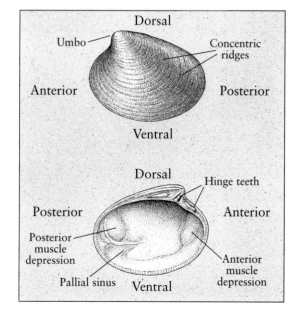

Typical bivalve shell, external view above, interior below. The specimen shown here is *Humilaria kennerleyi*, collected at Friday Harbor, Washington, United States. (Illustration by Roberto Osti, based on an illustration by Janice C. Fong.)

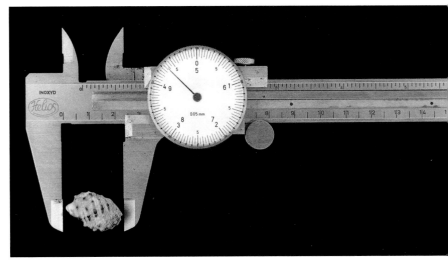

Dial calipers modified to enable the author to measure shells. The specimen being measured is *Plicopurpura columellaris,* collected from high-shore, wave-exposed rocks near Todos Santos, Baja California Sur, Mexico, on April 8, 1995. (Photograph by Mary Graziose.)

Volema pyrum, 61.8 mm long, collected from mud near Mahajunga, Madagascar, on July 6,1972; apertural *(left)* and dorsal views. The shell sustained two grave injuries that were later repaired by the snail. The first injury is visible on the shell whorl just to the left of the aperture. The second, which occurred later, appears on the dorsal side of the shell *(right).* (Photograph by Mary Graziose.)

Muricanthus radix, 64.7 mm long. This specimen was collected in May, 1976 at Playa Venado, Panama, from sand near rocks. (Photograph by Mary Graziose.)

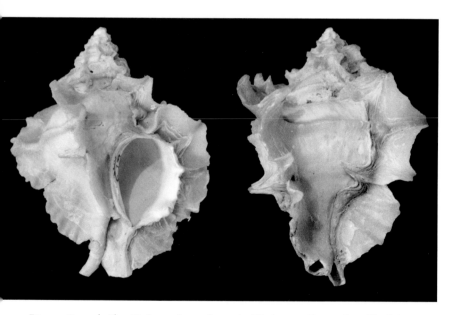

Pterorytis umbrifer, 52.0 mm long, from the Yorktown Formation, Virginia. This snail lived some three million years ago during the Pliocene epoch. A long spine protrudes toward the viewer from the shell aperture's outer lip *(left)*. A similar spine developed in earlier stages whenever growth temporarily ceased; one of these can be seen along the extreme left margin of the dorsal *(right)* view of the shell. (Photograph by Mary Graziose.)

The author with his daughter, Hermine, and wife, Edith.
(Photograph by Gerry Gropp.)

for hermit crabs. On the reef flats of Guam, these "dead" shells often outnumbered living snails by four to one. Provided the "dead" shells accurately reflected the frequencies of such causes of death as breakage and drilling, I could examine large samples not only from Guam but from other parts of the world as well, to estimate the importance of shell breakage as a cause of death.

Field sampling soon confirmed what I had suspected. For small species such as *Strombus gibberulus* and *Cerithium columna,* shell breakage accounted for most deaths, whereas for larger, more heavily fortified snails such as *Drupa morum,* most causes of death left the shell intact.

The tantalizing evidence we had gathered about the capabilities of the crabs and the defenses of shells was only the beginning. Guam's reefs and shores support a wide variety of predators—fish, mantis shrimps that hammer or spear their prey, carnivorous snails that drill into their victims or invade through the aperture, and even some small sand-dwelling sea stars that swallow their prey whole—that might be just as important as crabs are to the lives and deaths of molluscs. What were their capabilities, and how had molluscs responded to them evolutionarily? Moreover, we needed to find ways of making direct comparisons between Guam and other tropical sites with respect to the defense mechanisms of shells and the performance of predators.

A promising if not wholly satisfactory approach occurred to me late one evening back in College Park in Maryland. I knew that crabs were not terribly fussy eaters. Whether a shell contained the mollusc that built it or a hermit crab that occupied the shell after the death of its maker seemed to make little difference to a hungry crab. Why not take advantage of this indiscriminate behavior? I could bring freshly collected empty shells from other parts of the world to Guam, reinhabit them with local hermit crabs, and offer them along with local shells to individual crabs.

Shells for the upcoming experiment would come from the north coast of Jamaica. As Edith and I gathered the requisite hermit housing from the shores around Discovery Bay during the winter holidays, we also took samples of "dead" shells. Less than a quarter of these shells showed signs of lethal breakage, whereas in Guam, many samples contained broken shells at frequencies of 50 percent or more.

One more of my shell-collector's deep-seated prejudices had to go. Sometime in the early 1960s, my brother Arie had given me a specimen of the lettered cone *(Conus litteratus)* from the Philippines as a special birthday present. I knew that he had gone to considerable trouble to buy it at the American Museum of Natural History in New York, and so it was with a sinking feeling that I realized the specimen was badly flawed. Running in a jagged line from one end of the shell to the other was a scar, a rough, ugly welt rudely interrupting the otherwise smooth contour of the shell's pleasing conical shape. I tried not to show my disappointment, but I could never like the specimen as I did all the other shells in my growing collection. The scar was a blemish, an insult that some careless human collector must have inflicted on the specimen. In Discovery Bay, I began to appreciate its significance. It occurred to me that, if shell-breaking predators were responsible evolutionarily for bringing about the traits that made tropical shells such impressively strong fortresses, then their attacks would often be preserved on the shell as repaired damage. The welt on the lettered cone was not the work of a person, but the record of an encounter with a creature that damaged the shell at an earlier stage of growth. The living snail suffered damage at the shell's thin lip, which through regeneration and subsequent growth was mended. The scar was an entry in the skeletal diary of the snail's eventful life. If all attacks on shells by predators resulted in the prey's death, features that enhance a shell's resistance against attacks could not evolve. Survival would instead be contingent on a snail-damaging predator, and a heavily fortified shell would provide no survival advantage.

Once I recognized the evolutionary importance of scarred shells, I encountered repaired damage everywhere I looked. In many populations of auger shells, for example, nearly every individual's shell was marked by one or more long jagged scars recording the unsuccessful efforts of predatory crabs. I counted thirteen such scars on one specimen of the Hawaiian auger shell *Terebra gouldi,* which I turned up in a collection from the 1860s at Harvard's Museum of Comparative Zoology. Auger shells are especially well adapted to withstand the attacks of box crabs, which try to reach the prey's edible soft parts by cutting or peeling the shell in a spiral direction starting at the prey shell's thin lip. The soft parts withdraw so deeply into the long, narrow shell that box crabs often given up before they have cut far enough to reach them. I have seen living auger shells with scars extending one or even two whole turns, indicating that the shell is able to sustain extraordinary damage while still serving as effective protection for the snail within.

Snails in which the rim of the shell opening is heavily reinforced do not suffer lip damage that requires repair, but even here careful inspection often reveals the scars of battle. Assaults on *Drupa morum,* whose thick knobby shell exemplifies the most extreme armament in the snail world, are evident from the presence of broken knobs on the shell's exterior.

We returned to the western Pacific for a second summer of laboratory and fieldwork in May 1975. As we hoped, the crabs treated the foreign shells as they would any other. Once the predators were alerted to the presence of prey, their instruments swung into action. The minor claw grasped and turned the shell while the major claw tested and squeezed in a slow tempo. While I took notes, Edith timed and intently observed each movement. Minutes, often up to an hour passed until that final satisfying crushendo, when at last the edible morsel within was sufficiently exposed to be eaten.

Most of the Jamaican shells fared badly. The large topshell, *Cittarium pica,* not only had a vulnerably large aperture and a thin unbuttressed outer lip, but on its underside the shell bore a steep-sided funnel-shaped cavity, or umbilicus. A crab's claw could hook over the edge of this cavity so that, as its fingers closed, the shell split in half. The Guamanian topshells of the genus *Trochus* were more tightly constructed and had a smaller opening and a much less vulnerable basal cavity.

During the summer of 1975, we were joined in our laboratory work in Guam and in our later field studies in Palau by Rick Dickinson, a soft-spoken Stanford graduate working toward a master's degree at Guam under Lu Eldredge's direction. Despite a brush with cancer several years earlier, Rick personified hope and optimism. He observed, with dismay and disbelief, that I had managed to avoid learning how to swim. He ignored my lame excuses and promptly embarked on a vigorous training program, strongly encouraged and assisted by Lu's wife Jo. "You should not go to Palau without knowing how to swim," Rick said, and of course he was right. In fact, as we prepared one afternoon to return from a site on Palau's outer reef, the incoming tide had floated the boat to deeper water, and we were compelled to swim to it. What a lasting pity it is that this bright, level-headed, and personable young man died just two and a half years later of an unrelated rare form of cancer.

Meanwhile, the Palauan reefs and inshore habitats richly rewarded us with data and insights into molluscan death. Evidence of shell breakage was everywhere. Large cowries with their tops broken away lay alongside turban snails and top-shells with their spires sheared away. Cones and conchs had part of their outer lips cut away, usually enough for predators to have killed the original snail. Yet these shells, and almost all other damaged and intact "dead" shells, were still part of the ecosystem, for hermit crabs eagerly sought them as housing. Such flawed shells were in great demand, and there was clearly a very high premium on objects that could furnish protection for the crabs' soft abdomens.

If the shells of western Pacific Ocean and Indian Ocean snails were on the average architecturally more resistant to breakage than those in other parts of the tropics, could a corresponding difference in the power of crab claws be demonstrated? We chose an indirect way to find out by measuring the dimensions of claws of species from different parts of the world and calculating the mechanical advantage of the claw's movable finger. A crab's claw is actually a modified limb consisting of several segments joined by tight hinges that allow rotation of motion only about a single axis. It squeezes an item between two fingers, a movable upper one—the terminal segment of the limb—and a fixed lower one, which is an extension of the next to the last segment. The massive claw-closing muscle, located in the next to the last segment, causes the upper finger to rotate on its heavily reinforced hinge toward the fixed finger. A high, thick claw would indicate a claw-closing muscle of great power. Slender claws, by contrast, should exert weak forces or deliver short bursts rather than sustained muscle contractions.

Edith and I once again took advantage of the Smithsonian's crustacean collection, but this time we sampled it systematically and measured hundreds of specimens. There was also a superb collection at Leiden's Rijksmuseum voor Natuurlijke Historie (since blandly renamed Nationaal

Museum voor Natuurlijke Historie) in Holland. I could combine work in this museum with visits to my parents in Diemen, a suburb of Amsterdam, and to Arie and his family in Oudewater, a charming ancient town east of Gouda, where Arie managed the local NMB bank branch.

The crustacean collection at Leiden was housed in a peculiar building on the Raamsteg, in which the upper floors consisted of open iron grates. The alcohol that inevitably dripped from the preserved crabs drizzled onto the floors below. The molluscan collection was also stored in a room with this outrageously impractical floor. I have often wondered how many specimens in addition to the one or two I accidentally dropped found their way to street level.

Lipke B. Holthuis, the highly dignified curator of Leiden's crustacean collection, was one of those rare individuals ideally suited to his task. Besides an encyclopedic knowledge of every aspect of the biology and classification of his group, he possessed an extraordinary eye for the features that taxonomists use to distinguish among species. His papers, written in at least five languages, were peppered with asides about expeditions and accounts of natural historians of the past, for Holthuis was an authority on the history as well as the science of his field.

In Leiden, our measurements confirmed the predicted pattern. Two Indo-Pacific species of *Carpilius* had relatively thicker, higher claws than did the single Caribbean species, *Carpilius corallinus*. The genus *Eriphia* showed a similar interoceanic pattern, but with the additional feature that the single eastern Pacific species had claws whose relative dimensions lay between those of its Indo–West Pacific and Atlantic relatives. Moreover, tropical shell-crushing crabs generally had relatively larger, fatter claws with more specialized crushing teeth than did their temperate-zone relatives.

The results of this crab work, published in *Systematic Zoology* in 1977, were not universally appreciated. In June of the following year, Kenneth L. Heck, then of Florida State University, told me on a chance encounter at the Smithsonian that he and his colleagues Lawrence G. Abele and Daniel Simberloff were planning to write a piece that would criticize my 1977 paper. The Florida State group had become almost notorious for their sharp attacks on ecologists whose inferences of competition were considered by the group to be outright erroneous. I offered to send them my unpublished raw data, but Ken felt this would be unnecessary—just yet.

I had nearly forgotten about this conversation when Larry Abele called me in late September 1979. Why, he asked indignantly, had I ignored his request for my raw data? Why had I failed to respond to a manuscript he had submitted with Heck and Simberloff to *Systematic Zoology* in May? I was greatly surprised. It would indeed have been unpardonable of me not to have given the Florida State group my data, and stupid not to have commented on their manuscript which criticized my work; but the truth is that I had neither their request nor their manuscript. If they had sent their work in May, chances are it was lost in the mail, for Edith and I were underway from Guam via Manila and Sydney to meet the *Alpha Helix* at Darwin. In any case, I sent Abele my data, and he sent the manuscript, which was still under review at *Systematic Zoology*.

When I read the critique, I was disturbed by some of the strong language of condemnation. I disagreed with some of the complaints about methodology, but others of their points were justified, I felt, and I wrote to the editors of *Systematic Zoology* saying that the paper should be published, perhaps with a softened, less vitriolic tone. *Systematic Zoology* eventually rejected the paper, but suggested that the Florida group resubmit it, taking into account the reviewers' criticisms.

There matters stood until September of 1980, when Daniel Simberloff was scheduled to give a seminar in the zoology department at Maryland. Despite the Florida group's confrontational history, I approached Simberloff with a collaborative proposal. Could we, I suggested, write a joint paper in which the Florida State group could outline its objections and I would rebut or agree where appropriate? The dispute could be aired in a single, coherent, nonconfrontational account that readers could follow easily. The more traditional way of dealing with disagreements was to publish the critique and the rebuttal as separate papers, often appearing in different issues of the journal, but that format would enable the two sides to talk past each other and confuse rather than illuminate the debate.

A few weeks later, Larry Abele called me with a proposal of his own. If I were willing to come to Tallahassee to give a seminar, he and my other critics would be delighted to hammer together a single, joint paper with me. I could stay with Abele and his family and take the opportunity to look at some local salt marshes and freshwater springs. I accepted his offer.

From this point on, it was smooth sailing. From my perspective at least, the exercise of talking and writing with my critics was a highly constructive act from which I learned a great deal, and as a result, I gained a lasting respect for each of my three coauthors. The paper was published in *Systematic Zoology* in late 1981.

I tell this story because it illustrates how scientists work and publish their results today. It could have ended much less happily. Yet another angry exchange would have found its way into the scientific literature, to be ignored or dismissed as a tempest in a teapot. The admiration and respect that now prevails among this group of authors would not have been achieved. I believe science is not well served by the adversarial ethic that prevails in courts of law, and as an editor of journals I often tried to bring the two sides in a dispute together.

The kind of research Edith and I were conducting takes time, and it had to be done on location in faraway places. We had to observe and take care of animals every day, and weeks or even months might go by before we had enough data to be analyzed meaningfully. The normal academic year provides this kind of uninterrupted opportunity only during summers. I found the work so exciting that I wanted to devote more time to it. Accordingly, I applied for a John Simon Guggenheim Memorial Foundation fellowship in October 1974. Compared to the long, detailed grant proposals demanded by the National Science Foundation, the two-page statement that the Guggenheim Foundation was looking for required little effort on my part. If I were lucky enough to be among the 10 percent of successful applicants, I would have some research funds and, more important, a whole academic year to devote myself to my studies of molluscs and their predators.

One part of the world that continued to hold a special fascination for me was Panama. Shells of rocky-shore snails in the eastern Pacific seemed to be architecturally more specialized as breakage-resistant structures than were their nearby Caribbean counterparts, and the eastern Pacific species of *Eriphia* had relatively larger claws than did the West Indian member of the genus. I was therefore anxious to study crabs and snails on the two coasts of the isthmus of Panama.

The Guggenheim Fellowship for the 1975–1976 academic year enabled me to begin work in Panama by April, a month when I would normally be caught up in teaching introductory zoology.This is the hottest time of the year in Panama, when the smoke from burning grasslands at the end of the dry season is carried far and wide on the strong persistent trade winds. Upon our arrival, Edith and I learned that a certain

Allison R. Palmer from the University of Washington was expected soon, ready to work on the same problem we were. My considerable apprehension was immediately and permanently put to rest when Rich Palmer appeared.

Besides being a man of good humor and goodwill, Palmer was a first-rate experimentalist. Though only a graduate student, he already possessed formidable skills in statistics, shell measurement, and experimental design. I have never known a more incisive, hardworking, careful, or dedicated experimental scientist. Palmer has an intensely creative mind that asks intriguing questions and wrings important lessons from seemingly esoteric and arcane phenomena that others would ignore. His work on fluctuating asymmetry—small inconsistent differences between left and right parts of a more or less bilaterally symmetrical animal—the energetics of calcification in shell-building, and the effects of crabs on the growth and thickness of shells of their prey species has set a very high standard of rigor that is rarely surpassed.

While Edith and I worked with crabs and a spiny lobster as predators, Palmer set his sights on spiny puffers. We had come to know these slow, large-eyed, thorny fish well in Guam. Their massive upper and lower jaw plates close like a vise on snails and hermit crabs, whose shells shatter into innumerable pieces under the pressure. In one puffer's stomach, we recovered the remains of fifty-eight prey, including thirty-eight individuals of *Drupa rubusidaeus* (the raspberry drupe), a heavily fortified snail species. The same species of puffer, *Diodon hystrix,* also frequented the Bay of Panama. Palmer discovered that captive individuals readily learned to take snail prey from his hand. Indeed, the fish were sometimes a little too eager, for they snapped menacingly close to the hand that fed them. Palmer soon demonstrated that the presence of large tubercles made shells effectively bigger and thus prevented the puffers from fitting the prey into the vise. With the tubercles artificially removed and the shell therefore made smaller, individual snails previously too large to be eaten were summarily crushed by the voracious puffers.

Another great experimentalist soon joined the effort. In February 1976, I received for review a manuscript about temperature relationships in *Nucella lamellosa*, a common dogwhelk in the Pacific Northwest, from the editor of *Ecology*. I knew neither of the authors but surmised that one of them, Mark D. Bertness, was either an undergraduate or a master's degree candidate at what was then still known as Western Washington State College in Bellingham. On the strength of this hunch, I wrote Bertness a letter expressing my great pleasure with his paper and asking if by any chance he had ambitions to pursue a doctorate. He would be welcome to work with me if he did.

To my surprise and everlasting gratitude, Mark said yes. Undistinguished performance on the Graduate Record Examination and in his undergraduate courses had unnecessarily and sadly doomed his applications to more prestigious graduate schools, but I argued to Maryland's admissions committee that the *Ecology* manuscript was proof enough that Mark was a gifted experimenter with a first-rate intellect.

There was a palpable intensity, and unstoppable drive, and an irreverent streak in Mark. Under the wild crop of curly blond hair lurked a formidable unruly mind, from which came inspired experimental designs and an uncompromising work ethic. By the end of his first year at Maryland, he was ready to take the oral comprehensive examination and to embark on an ambitious research project on hermit crabs on the two coasts of Panama.

As soon as he arrived in Panama in June 1977, Mark plunged into experiments. Taking advantage of the enormous abundance of hermit crabs and the ready availability of several geometrically contrasting shell types, he quickly confirmed by experiment the role of large size, strong sculpture, thick shells, and narrow or small apertures in protecting snails and hermit crabs from predators. These catalogued benefits of protection, however, came at a cost. Good fortresses contained little space for eggs, which the hermit crab carries on her soft abdomen inside the shell. They also diminish the ability of hermits to withstand the heat and aridity of the upper shore. When crabs compete for shells, the kind of shell they prefer depends on which function of the shell confers the greatest advantage. Species low on the shore, where predation is common, may prize good fortresses, whereas species living between tidemarks tend to choose roomy shells in which crabs find effective shelter from the weather.

By the time Mark had published the last of eleven papers on the Panamanian work, nine of which comprised his dissertation, he had marshalled a compelling case for the hypothesis that predation is more intense on the Pacific than on the Atlantic side of the Panamanian isthmus, and that adaptation to it was correspondingly greater. On the basis of this work, Mark was offered a position at Brown University, where he now holds the rank of professor. In the years since his thesis, he has become one of the leading marine ecologists of our day, with important contributions on the biology of salt marshes and on the larval recruitment of shore invertebrates.

In the spring of 1975, shortly after being awarded the Guggenheim fellowship, I received a letter from William Bennett at Harvard University Press. Would I, he inquired, be interested in writing a book? Initially, I dismissed the idea without giving it serious thought. I was too young, and in any case I didn't have enough to say to fill a book. As the results of the summer's crab experiments and the surveys of broken shells piled up, however, I reconsidered the matter and wrote Bennett that I might take up his invitation. I wanted to write a book that explored the geography of shapes instead of the more conventional topic of the biogeography of names. Ever since the day in 1968 in Hawaii when I realized that snails from the tropical Pacific looked different from those in the Caribbean, I understood that simply knowing the geographic limits of species distributions was not enough. The names had to come alive. They belonged to individual plants and animals that interacted with other individuals; they had distinctive architectures, behaviors, and histories. Lurking beneath the details of place names, species names, and organic shapes were intriguing patterns and principles that provided important clues about how factors important to evolution, such as competition and predation, varied from place to place.

For two years, I labored mightily on the nine chapters that would comprise *Biogeography and Adaptation: Patterns of Marine Life*. Writing nine chapters proved to be far more work than writing the equivalent number of papers. I couldn't just report the geographical patterns I observed in the snails and clams that particularly interested me. I had to think about them and to place them in the context of the work of others. Did my marine perspective have any relevance for geographical patterns on land or in fresh water? How did my results for molluscs and crabs compare with patterns in corals, seaweeds, and fish? By the time I finished the first draft in June 1977, I was exhausted and ready to put some distance between myself and the manuscript. Edith and I went off to Curaçao and Venezuela for some much-needed and interesting fieldwork.

Writing a book is a bit like having a child. My anxiety rose to unprecedented heights as I worried about who the reviewers of the manuscript would be and what they would say. Had I overlooked something obvious? Were there dreadful flaws in the arguments? Was I so close to the subject that no outsider could follow my prose?

My reviewers, Rich Palmer and his advisor Robert T. Paine at the University of Washington, were kind to me. Although they offered plenty of suggestions for improvement, which I eagerly incorporated, they judged the manuscript to be both readable and informative. I sent the revised manuscript, which was at least a third longer than the first draft, back to Bennett in the early autumn.

The publication of the book in 1978 was like the birth of a child both in bringing to a close one phase of uncertainty and scrutiny and in ushering in another. Books, unlike scientific papers in journals, are publicly reviewed, and authors are powerless to answer their critics. I had read enough book reviews in the pages of *Nature* and *Science* to know that critics could be savage and cruel in their condemnations. One reviewer of an encyclopedia remarked that the book should never have been allowed to leave the warehouse. Another opined that the volume he was asked to review was the worst thing he had ever read on the subject. Some of these reviews might have been on target, but others were tragically off the mark. I was haunted by the reception of Henry Horn's book, *The Geometry of Trees* (Princeton University Press, 1971), a brilliantly original and enormously stimulating piece of ecological, biomechanical, and mathematical scholarship that was roundly condemned as second-rate by a highly respected but deeply misguided commentator in *Science*.

Edith and I were in Guam early in 1979 when Jim Marsh, one of the scientists at the marine laboratory, walked in with the latest copy of *Science*. Knowing that we were always eager to read the weekly magazine, he handed it to us unopened and unread in the knowledge that we would return it promptly. Edith gasped as she scanned the table of contents. "Here's a review of your book," she said, with a touch of fear in her voice. If I could hardly bear to listen to the review, Edith found it even more trying to read it aloud to me. Happily, David Woodruff, an evolutionary biologist now at the University of California, San Diego, was very pleased with it. Just as gratifyingly, so was Philip Morrison in *Scientific American.*

Chapter 12

THE STRUGGLE OF TIME

The metaphor emerging in my mind from our tropical work was that of a standoff in an arms race between two opposing forces. On one side were the predators, outfitted with powerful, heavily reinforced claws or jaws; on the other were the prey, well defended in sturdy but elegant castles of calcium carbonate. In the mid 1970s, I thought of this predator-prey tug-of-war as a typical example of coevolution, a process in which two interactors—host and guest, plant and pollinator, a predator and its victim, or even two competitors—evolve in response to each other. Although Charles Darwin and other nineteenth-century biologists were already familiar with such reciprocal adaptation, the concept of coevolution was given new life in 1964 with the publication in *Evolution* of Paul H. Ehrlich and Peter H. Raven's classic paper on butterflies and plants. Noting that many groups of butterflies feed only on certain families of plants, Ehrlich and Raven held that successive bouts of diversification of the two species were made possible when one side gained a temporary advantage over the other. The plants might "invent" a new chemical defense, or the insects might evolve a new way of neutralizing the plants' toxins. The group under siege would then evolve countermeasures, which would tilt the balance temporarily in its favor.

The regional variations in shell architecture now appeared in a new light. Predator-prey coevolution might have proceeded furthest in the tropical Pacific and Indian oceans, to an intermediate level in the tropical Atlantic, and to only a very modest degree in the colder waters of the temperate and polar zones. But why? What factors would propel interacting species to high levels of specialization in some places and not in others? I was back to a version of the same question I first asked in fourth grade: why are tropical shells smooth and polished and cold-water ones chalky?

It seemed to me that two equally important factors were at work. The first was the dependence of most chemical reactions and biochemical processes on temperature. As a general rule of thumb, bodily functions double in speed with a rise in temperature of 10 degrees Celsius. Moreover, calcium carbonate—the mineral of which shells are built—forms more readily in warm water than in cold water. Animals that live in the higher temperatures of the tropics can attain higher rates of metabolism, locomotion, shell-making, nerve conduction, and other essential biological functions. In other words, specializations that require high performance levels are energetically more attainable, and are therefore more achievable through natural selection, under warm-water conditions. The second factor was history. The creatures and interactions we observe today are shaped by more than present-day processes; they are the products of time, and time leaves its indelible mark on all aspects of ecology, physiology, diversity, distribution, and organic architecture. The excellent and relatively well studied fossil record of shell-bearing animals provided an unparalleled opportunity for probing this history, one unavailable to biologists working with less easily preserved creatures.

The shallow-water parts of the tropical Pacific are characterized not only by shells with elaborate antipredatory defenses, but also by an apparent prohibition against features that make shells vulnerable to breakage. These features—a wide opening, thin shell, unreinforced apertural rim, and low overlap between successive coils of the shell—are absent in shallow-water snails from the warm Pacific and Indian oceans except under unusual circumstances, but they occur frequently in snails found in regions or habitats where the danger of breakage is low. Such ecological refuges from breakage are found in the cold waters of the deep sea and at high latitudes, as well as on the surfaces of seaweeds, in deep recesses and cavities under boulders and beneath corals in reefs, and in freshwater rivers and lakes.

In November 1974, I finally perceived the connection between this pattern of architectural permissiveness in ecological refuges and the shapes of snails that lived during the remote past. From the time I began work on my junior paper I had been intrigued by snails with unusual patterns of shell coiling. Today, most snails have right-handed shells. Beginning near the tip or apex of the shell, the aperture traces a clockwise spiral as it expands and as the shell grows. The opening appears on the right side of the shell when it faces the observer and when the shell is held with the apex pointing up. A few living snails have left-handed shells, in which the aperture traces a counterclockwise path as the shell grows. Fewer still are coiled in a plane. Shells of this type are generally much less well buttressed from within than are either right-handed or left-handed shells, because critical sectors of the shell wall are unsupported by overlapping portions of previous shell coils. They are therefore potentially vulnerable to catastrophic breakage and are not well represented in regions and habitats where shell-breaking predators abound. Instead, shells that coil in a plane occur mainly among land and freshwater snails, as well as in a few unusual marine habitats such as deep-sea muds and the surfaces of seaweeds. During the Paleozoic era (540 to 250 million years ago), however, snails of this geometry were large and conspicuous members of warm-water communities in shallow seas. If these ancient snails architecturally resembled snails living in rivers and lakes at temperate latitudes today, then perhaps the predators that fed on them were relatively weak, as are freshwater shell-breaking predators today. Marine warm-water shells with a more modern predation-resistant architecture—a narrow opening, reinforced rim, and tight coiling— did not appear in significant numbers until 150 to 200 million years ago during the Jurassic period of the Mesozoic era. Their appearance coincided with the waning of the more archaic types with broad unreinforced openings and relatively loosely coiled shells. As this architectural shift was taking place, new predators with more powerful instruments for breaking and entering shells appeared. There were crabs with crushing and probing claws, predatory snails that drilled their victims or entered by way of the prey's aperture, sea stars that pried open the tightly closing valves of clams with flexible arms, and fishes with specialized cutting teeth in massive jaws. I published a preliminary paper on this topic in *Nature* in 1975.

It soon became obvious that armored molluscs and their predators were only a small part of a much larger and more interesting story. Many patterns and observations that previously made little sense now fell together into a coherent framework: shallow-water marine communities underwent a profound general reorganization beginning perhaps 200 million years ago. The cascade of events involved changes in architecture, the invasion of less thoroughly exploited habitats, an increase in the number of species, and increased reliance on methods of predation and consumption that demanded the use of abundant metabolic energy. For example, animals such as sea urchins, which during the Paleozoic had browsed the tips and other softer portions of their algal food, added scraping and rock-gouging to their feeding repertoire during the Mesozoic era. Marine sandy and muddy bottoms, which had been churned up only in the topmost layers by burrowing animals during most of the Paleozoic era, were becoming occupied by a wide assortment of deep-burrowing sea urchins, sea cucumbers, lobsters, snails, clams, and other animals. And there was more.

During the summer of 1976, Edith and I spent eight delightful weeks at the Friday Harbor Laboratories on San Juan Island in the state of Washington, where I was teaching a graduate course with Alan J. Kohn. The University of Pennsylvania's Charles W. Thayer, whom I knew well from my graduate student days at Yale, was carrying out interesting work with brachiopods, clam look-alikes whose bivalved shells are among the most abundant fossil remains from the Paleozoic. These animals, he discovered, have lower rates of metabolism and slower body functions than bivalved molluscs, which did not begin to establish strong numerical dominance over brachiopods until early Cretaceous time, about 130 million years ago. Moreover, Thayer showed that brachiopods, which normally live attached to objects on the seafloor by a muscular stalk, cannot reattach once they are dislodged. Bivalves fixed to the seabed by a tangle of tough threads, however, can readily do so. All these observations pointed to increased biological activity on, above, and below the seafloor beginning in the middle of the Mesozoic era. In a paper in *Paleobiology* in 1977, I rather provocatively termed this reorganization the "Mesozoic marine revolution."

Once again, I faced the challenge of testing aspects of my hypothesis. Besides documenting the pattern of shell architecture through time in more detail, I needed evidence of a kind not previously considered in formulating my ideas. If predators became more common, or if shells became relatively better fortresses, the frequency of scars should have increased through time. Testing this prediction required exceptionally good fossil material in which the details of external ornament were well preserved. Few Paleozoic and early Mesozoic fossil assemblages of snails met this stringent criterion.

Porter Kier, an authority on fossil sea urchins who also happened to be director of the Smithsonian's natural history museum, pointed to the remarkably diverse fauna of the late Triassic St. Cassian Formation of northern Italy. The quickest way to study this beautiful material was to visit Rinaldo Zardini, a seventy-seven-year-old amateur who had built his own museum in Cortina d'Ampezzo to house his extensive Triassic collections. Zardini, an exceptionally energetic man, was thrilled when Edith and I visited him in June 1980. Between long walks through the Alpine woods and meadows, we measured and examined hundreds of fossils. With Zardini, Edith and I published a paper in the *Journal of Paleontology* in 1982 reporting our finding that these Triassic snails showed very few repair scars.

The study of a late Paleozoic assemblage proved to be a much more ambitious undertaking. David E. Schindel, then at Yale, had amassed a very large collection of exquisitely preserved snails from several formations of Late Carboniferous age (about 300 million years ago) in central Texas. He, Edith, and I spent days cleaning, measuring, and inspecting thousands of specimens at Yale's Peabody Museum. Although the frequency of scars was greater than in Zardini's Triassic material, it was still generally low compared to the frequency of scars in the Late Cretaceous snails that my assistant Bettina and I were studying at the Smithsonian, and compared to my own material from the late Miocene (about 5 to 8 million years ago) of Panama and from various tropical sites in the modern ocean. We summarized the data on healed injuries through time in a paper in *Science* in 1981.

Even among molluscs, there was more to the Mesozoic revolution than just armor. Escape was another evolutionary option, available especially to molluscs whose main predators are relatively slow. Their only way to escape from fast predators such as crustaceans, fish, and humans is to drop from a rock into a nearby crevice. I witnessed a spectacular instance of this behavior in Madagascar in 1972. As Edith and I walked along the boulder shore on Nosy-Be, an island in the northwestern part of the country, hundreds of *Nerita doreyana* snails lost their grip on the stones and tumbled noisily into interstices as we approached. I suspect the snails detected our footsteps through vibrations. When molluscs are dealing with other snails or sea stars as enemies, however, quick movement—burrowing, crawling, swimming, and even jumping—is necessary. Steven Stanley had already learned a great deal about the shell features that enable clams to move rapidly, but similar work on snails was limited. Edith and I therefore decided to return to Guam in 1984, this time with our two-year-old daughter Hermine, to study snail burrowing. Placing snails of different sizes, shapes, and surface sculptures on sand in an aquarium, we conducted time trials to ascertain how rapidly the animals buried themselves completely. Two years later, Bettina and I repeated this protocol with sand-dwelling snails in Panama.

Not surprisingly, snails with a large muscular foot and with a smooth shell surface rapidly disappeared beneath the sand, whereas those with a small foot and a rough shell exterior burrowed slowly or not at all. A streamlined shell shape was no guarantee of fast burrowing, because streamlining often results in a narrowing of the shell opening, which in turn correlates with a small, weak foot.

The designers of tanks, automobiles, and aircraft have long appreciated the principle that armor and speed are incompatible. If snails in the western Pacific and Indian oceans excelled in resistance to breakage, they should show only a modest adaptation for speed. This is indeed the case for many individual species; but for burrowers, the Indo-Pacific was as conducive to the evolution of traits that made high speed possible as it was to the evolution of armor. Often, this was achieved by means of a bigger engine, the snail's foot. Olive shells, for example, can disappear beneath the surface of the sand in less than a minute by virtue of their huge foot, which envelops the shell. When threatened, however, the snail can deflate the foot and withdraw the entire body into the smooth, sleek, cylindrical shell, whose narrow aperture makes entry by predators difficult and time-consuming. Elsewhere in the tropics, shell traits that enhance the ability to burrow are less common, and in the temperate zones they are decidedly rare. Inspection of fossils showed that most Paleozoic and early Mesozoic snails lacked burrowing-related features and that the nimblest living groups had origins within the last 30 to 40 million years of our own Cenozoic era. Through their evolutionary history, therefore, snails and other groups of marine animals specialized along many pathways—resistance, aggression, speed, toxicity, and cryptic coloration—in response to predators and other enemies.

With so many predators involved and so many evolutionary directions being taken, I came to realize that coevolution is an inadequate model for explaining a cascade of events as complex and global as the Mesozoic marine revolution. Instead, I began to think of the Mesozoic episode of large-scale change, and of comparable reorganizations during the preceding Paleozoic era, as manifestations of a process of escalation, in which selection due to enemies drives much of the adaptation we see. In this conception, evolutionary control comes mainly from competitors and predators and much less from food organisms. Snails adapt to shell-breakers by fleeing or by resisting their predators; the predators, in turn, respond more to their own enemies—other predators and competitors— than they do to their passive prey. In special cases, there could be two-way (or reciprocal) coevolution if two interacting species mutually benefited or mutually harmed each other, but such coevolution would represent a special form of escalation in which the interactors were able to respond to each other. Most interactions in nature are evolutionarily lopsided. Species A can significantly affect the evolution of species B, but species B often has a minimal effect on species A. Even the forging of alliances between species can be understood in the framework of escalation, because such partnerships often result in the evolution of potent competitors whose evolutionary effects on other species can be profound. Thus the mutually beneficial relationship between a host coral and guest single-celled plants produces an organism capable of rapid growth, effective defense against consumers, and effective competition with neighbors. Some species, of course, can adapt to the evolution of such potent partnerships; but others cannot, and they are relegated to regions or habitats where the risks from powerful competitors and consumers is low. Competition and consumption, in other words, dictate which phenotypic traits enhance survival and reproduction and which traits commit their bearers to death or to ecological banishment. As long as there are resources—food, shelter, mates, and allies—that are either themselves organisms or are under the control of organisms, there will be competition for those resources; and competition, provided species can respond to it evolutionarily, is the engine that drives adaptive evolution through a process of escalation between species and their enemies.

This model of escalation accounts for the directions in which evolution takes place—toward greater competitive ability, better defense, higher reproductive output, and occupation of marginal environments—but it is silent on the question that prompted me to study predation in the first place: why has escalation proceeded further in some regions and habitats than in others? In fact, should the process not be taking place everywhere and all the time?

This expectation was aptly captured by Lee Van Valen, a brilliant if eccentric biologist at the University of Chicago. As the first paper in his own journal *Evolutionary Theory,* he published "A New Evolutionary Law" in 1973. Taking a metaphor from Lewis Carroll's *Through the Looking Glass,* he proposed the Red Queen hypothesis, which holds that each species must "run in place" just to keep up with all the other species evolving with it. Such an interpretation was eminently reasonable for, as already well understood by Darwin, organisms are never perfectly adapted to their surroundings; they exhibit an evolutionary lag. In the late 1970s and early 1980s, I surveyed a scattered scientific literature for data on how often predators fail to identify, catch, and subdue their prey. Most predators, it turns out, fail during one or two of these steps of predation in more than 50 percent of attempts. Even the lion, so often portrayed as that most perfectly adapted of predators, fails to catch its prey in two out of every three tries. There is, in other words, almost always room for improvement. The inequities of predation and competition should fuel rapid, continuous adaptive evolution. But do they?

Evidence from an entirely different quarter was pointing to limitations on the ability of plants and animals to respond adaptively to changes in their environment. Paleontologists had long known that species, once they appeared in the fossil record, remained essentially unchanged in shape for long periods. Times of evolutionary change were brief by comparison. Alfred Fischer had pointed out this pattern in his course at Princeton in 1966. Niles Eldredge at the American Museum of Natural History recognized the phenomenon in trilobites in a paper published in *Evolution* in 1971. A year later, the pattern received a name, "punctuated equilibria," bestowed by Niles Eldredge of the American Museum of Natural History and Harvard paleontologist Stephen J. Gould. Times of evolutionary stability, or stasis, alternate with much briefer intervals of rapid change.

Controversy has swirled around the concept of punctuated equilibria ever since. How common is this pattern in the history of evolutionary lineages? Is adaptive change truly limited to short intervals, or does it occur gradually over thousands of generations? To what extent is the observed pattern of long-term stasis and short-term change an artifact of an incomplete, gap-filled fossil record? Although these and many other questions have not been fully answered, I believe most paleontologists have come to accept the premise that prolonged periods of an evolutionary status quo are very widespread among species. In other words, something is preventing species from adapting, from responding to agents of natural selection, in addition to the many adaptive shortcomings of individual plants and animals. Competition, predation, and mutual benefit cause natural selection, but that selection enforces the status quo more often than it results in an adaptive response.

With the perspective of punctuated equilibria in mind, I could now think of escalation in a new way. If adaptive stability, stability enforced by compromise among conflicting demands, rather than adaptive change is the norm in the evolutionary history of life, escalation between species and their enemies could proceed only when many species at the same time were released from the conditions that enforced the status

quo. Instead of asking why response to enemies leads to varying degrees of biological sophistication in different places or at
different times in Earth's history, I might more profitably inquire when, and under which conditions, escalation is possible.
What are the circumstances that prevent an adaptive response,
and when are such constraints lifted or altered? Is adaptive
change often too rapid for species to respond by means of natural selection, which involves a generation-to-generation
change in gene frequencies? Is there insufficient heritable variation on which the selection that leads to adaptation can act?
Does the way in which organisms are put together make some
directions of evolutionary change more likely to be followed
than others?

As I see it, all these possibilities play a role, but the most
important constraint on adaptive evolution is the environment
itself. The basic problem is that almost every adaptive response
comes with strings attached. Improvement in the performance
of any one function—defense, locomotion, or reproduction,
for example—usually entails a reduction in the performance
of another. Mark Bertness's experiments on hermit crabs in
Panama yielded some compelling examples of this trade-off
principle: higher fecundity meant slower growth, and an effective shell defense often compromised the ability of hermit crabs
to withstand the physical rigors of the tropical shore. An increase in an animal's speed usually comes with a reduction in
the weight, and therefore in the protective role, of an external
shell. Rapid growth often provides a competitive size advantage in animals, but it is achieved at the expense of sturdy
skeletal construction. In other words, few adaptive benefits are
universal. In some situations, such incompatibilities between
competing demands matter little. A crisis during which thousands of species become extinct, for example, might leave the
survivors in a world brimming with potential food and space.

With the everyday limitations prevailing during conventional times removed or relaxed, populations are free to expand, and improvements can be introduced with less drastic consequences for other critical body functions. Whether such unfettered population expansion is favorable to adaptation depends on the biological environment in which it takes place. If potent competitors and predators take part in the expansion, then enemy-related adaptations could readily evolve as the various species in a community expand and as they respond to their enemies through natural selection. If, however, extinction wiped out the more sophisticated competitors and consumers, as many studies of the great mass extinctions imply, then the recovery from crises may not be accompanied by escalation, because selection imposed by potent enemies will be initially weak.

Many of the attributes that make organisms formidable competitors and potent consumers are contingent on the availability of a plentiful supply of food and energy. Rapid growth, high speed, great endurance, high rates of photosynthesis, powerful musculature, and the performance of many life functions simultaneously all confer significant competitive, defensive, and reproductive advantages; but they all require a large per capita energy budget. Therefore, circumstances that enable individual organisms to capture and use more energy should be highly favorable to escalation. Warm climates make energy-intensive modes of life feasible; so does warm-bloodedness as found in mammals, birds, and many insects. Intervals of Earth's history during which the global climate warmed or tropical climatic belts expanded to higher latitudes should therefore have been especially conducive to the evolution of sophisticated defenses and competitive means, because many species can simultaneously and adaptively respond to their enemies by employing energy-intensive mechanisms or by entering into partnerships that themselves also demand high energy use.

My claim that adaptation, especially that of organisms to their enemies, constitutes a primary theme in evolution received a decidedly cool reception from many of my colleagues. When I presented some of my findings at scientific meetings, they were either dismissed as being unique to the evolutionary interactions between molluscs and their predators or ignored as the ramblings of an unrepentant adaptationist.

The decade during which my ideas were taking shape witnessed a concerted attack on the Darwinian view that natural selection and adaptation explain evolution as chronicled in the fossil record. The challenges issued from many quarters and profoundly affected thought throughout the disciplines of evolutionary biology, paleontology, and ecology.

One sprang from molecular biology. Comparisons of different species showed that the rate at which protein structure evolved by means of the substitution of one amino-acid component for another was roughly constant, or clocklike, rather than variable or irregular. To many observers, this constancy meant that most of the substitutions, or evolutionary changes, were neither beneficial nor deleterious, but were instead more or less neutral in their effects on the fitness of individual organisms. If evolution at the molecular level is often nonadaptive, a possibility that is eminently reasonable especially when considering changes in the primary genetic material, the same might be true for the more readily observed traits of form and behavior. Perhaps most evolutionary change is nonadaptive.

Many paleontologists found it easy to embrace such interpretations. Like the goods of human industry, organic architectures incorporate features that are best thought of as "fabricational noise," to use an apt term proposed by the German paleontologist Adolf Seilacher. The seam on a plastic cup, for example, does not make the cup a better receptacle for fluids, nor does it improve our ability to drink from the cup. Instead, it is a trace left by the process of manufacture. Similarly, the growth lines on the surface of a clamshell reflect an underlying pattern of growth and are not themselves adaptive features, although the pattern of growth and the size and spacing of the incremental lines are potentially subject to natural selection. On a more subtle level the pattern, or "program," by which an organism develops from the fertilized egg to adulthood may predispose the evolutionary lineage to which the organism belongs to evolve in one direction rather than in some other. Snails tend to be right-handed not because right-handedness is inherently superior to left-handedness, but because the basic asymmetrical structure of the coating of the unfertilized maternal egg predisposes any subsequent shell coiling to proceed in a clockwise rather than a counterclockwise direction. In other words, not every detail of an organism's architecture, physiology, or behavior is the product of adaptive evolution. Often, it is an evolutionary relic, a bit of harmless historical junk, that enables evolutionary scientists to infer pathways of descent from ancestral forms. If every trait were finely tailored to an organism's present circumstances, and if every trait were capable of adaptive modification in any beneficial direction, all traces of the evolutionary past would be erased.

The publication of Harvard biologist Edward O. Wilson's *Sociobiology* in 1975 crystallized simmering discontent into strident opposition to adaptationist interpretations in evolutionary biology. From his extensive research on social insects such as ants and bees and his careful reading of a vast literature on animal and human behavior, Wilson unabashedly explained many attributes of social behavior, including that of humans, as adaptations. Our own behavioral characteristics, Wilson argued, are under significant genetic control and have been inherited and modified from our primate ancestors. Stephen Gould and his Harvard colleague Richard C. Lewontin mounted an all-out attack on Wilson and on the field of socio-biology he helped to found. They accused Wilson and his fol-lowers of telling "just-so" stories, plausible explanatory ac-counts in which adaptation was assumed rather than rigorously demonstrated. The genetic basis of many traits—a necessary condition for adaptations—was unproved, Gould and Lewontin maintained; and the purported benefits of many be-havioral and social traits were imagined rather than factual. Although it was Wilson's embrace of the genetic basis of many human behavioral and social adaptations that sparked the loud and often very ugly debate about the role of adaptation in evo-lution, Gould and Lewontin were highly critical of the tacit ac-ceptance by biologists of adaptationist thinking generally. In 1979, these authors—one an eminent paleontologist, the other a leading population geneticist—crafted their attack on what they disparagingly labeled the "adaptationist programme" in a powerfully written essay in the *Proceedings of the Royal Society of London (Series B)*. Their piece, a model of clear writ-ing, deeply affected the thinking of biologists and paleontolo-gists around the world, and it was widely read and discussed even by scholars in such other disciplines as art history and lit-erary criticism.

Another challenge to the adaptationist viewpoint came from computer simulations of evolutionary patterns. In 1973, four authors—David Raup, Stephen Gould, Thomas Schopf, and Daniel Simberloff—published a landmark paper in the *Journal of Geology* showing that complex and realistic histo-ries of diversity within evolutionary branches (or clades) could be generated by the repeated application of a few simple rules. Over a given interval of time, the number of members of a clade could shrink by one, remain the same, or increase by one according to a random choice of fixed probability. Gould and Raup later carried out additional simulations showing that architectural patterns through time could likewise be generated repeatedly by applying rules without reference to any selective mechanism. The authors acknowledged that individual changes can still reflect adaptation or selective extinction, but the aggregate pattern, the overall behavior of the clade through time, obeyed only the laws of randomness. One could say by analogy that individual car crashes have unique causes, but the overall pattern of accidents does not reveal them and does not strongly depend on them.

At about the same time, several prominent evolutionary biologists and paleontologists—among them George Williams, Richard Lewontin, Stephen Gould, and Steven Stanley—were laying the groundwork for a view of evolution that dealt more effectively than did the prevailing Neodarwinian synthesis with the full range of the biological hierarchy. Classical adaptation and natural selection were expressed at the level of individual organisms. Yet there are more levels in the hierarchy of life— genes, cells, populations, species, and clades—where selective processes of sorting could well be at work. Stanley formalized this idea in a paper in the *Proceedings of the National Academy of Sciences* and elaborated his theory of species selection in his 1979 book, *Macroevolution, Pattern and Process*. The factors that controlled whether a lineage would split, remain unchanged, or become extinct might be quite different from those controlling the births and deaths of individuals. If an exquisitely adapted individual belongs to a lineage of large-bodied creatures, and if large body size makes a species more prone to extinction by virtue of smaller population size, as it does in mammals, then the fate of the purported adaptation is dictated more by the large-scale events that bring about extinction than by the everyday advantages individuals derive from it.

By 1985, Gould concluded in an essay in *Paleobiology* that adaptation at the level of individual organisms plays a subservient role in evolution to the factors controlling the birth and extinction of species. The mass extinctions had the greatest reverberations in the history of life, Gould maintained, because they eliminated large incumbent groups while providing renewed opportunities for the clades that happened to persist. He acknowledged that the snail-crab arms race I had documented might be the work of natural selection, but he saw this as an exception to the proposition that evolutionary trends arising from selection are short-term adornments on the larger patterns arising from nonadaptive sorting among clades.

These antiadaptationist sentiments were palpable at the Macroevolution Conference held in Chicago in October 1980. The meeting had been called to discuss the future course of the new field of paleobiology and to celebrate its accomplishments. Population geneticists such as Russell Lande clearly disavowed punctuated equilibria, but the attending members of the press accurately reported the overall strong endorsement of ideas conflicting with the Neodarwinian synthesis. During a reception one evening before formal proceedings got under way, I was approached by a young graduate student who, quite unprovoked, asked me in an accusatory tone, "Aren't you an adherent of the adaptationist programme?"

So I was, and so I remain. As a natural historian, I am impressed with the environmental rigors and challenges that individual organisms face in the unruly outdoors. If plants and animals do not resist or respond to the insults and opportunities in their surroundings, they will not survive and they cannot leave offspring. Adaptation may not be perfect, and it may be subject to compromise on all sides, but it is inescapable. This applies as much to clades as it does to individual plants and animals. There can be no question that emergent properties — population size, geographical range of a species, and the like — influence the probability that a population or species will give rise to daughter species or become extinct, but those properties are called *emergent* because they derive from the attributes of individuals and of the genes that encode them. As long as entities succeed or fail according to the heritable traits they possess, they will adapt no matter where they fall in life's hierarchy.

In a world that is increasingly urban, many biologists encounter plants and animals chiefly in such managed settings as gardens, pet stores, zoos, and laboratories. Lewontin's fruit flies in the laboratory live in an artificial universe, one ideally suited for the study of genetics and development and for experiments probing the power and limits of artificial selection. Darwin, too, could draw lessons about selection from his studies of animals under domestication. An appreciation of natural selection, however, requires observation and experimentation in the wild. If fewer scientists have that privilege, it is perhaps not surprising that the pivotal role of adaptation in evolution has come to be questioned.

It is only a small step from studying organisms in highly controlled environments to thinking of them or their fossil remains as lifeless entities more akin to stones or to molecules in an idealized gas than as interacting beings in a hostile world. Even the shells of my boyhood were objects of architecture, divorced from the environments in which their makers built them and depended on them. Not until I saw molluscs living in nature, and then connected their conditions of life with my terrestrial observations in New Jersey, did I gain some appreciation for function and adaptation.

Gould, who has been a good friend since our first meeting at Harvard in 1969, and whose work on land snails I have always admired greatly, has an especially interesting history. When he began to publish papers on the land snails of Bermuda in the late 1960s, Gould explored adaptive themes related to changes in shape as the snails grew in size. He argued, for example, that the size of the foot was excessively large in young individuals, and that progressive alteration in the shell's shape as the snail grew larger produced a foot of just the right size in the adult. Gould recognized very early that many shell traits change predictably according to the snail's growth pattern. Slow growth produces high-spired shells with a narrow umbilicus, whereas faster growth makes low shells with a wide umbilicus. Shell shape and even features of sculpture and color change because, as in most other animals, the growth rate slows with age.

Since the early 1970s, however, Gould has increasingly emphasized the importance of constraint in evolution. The grand themes of organic design—the genetic and organizational rules by which plants and animals develop from the egg to maturity—constrain morphology to vary in only a few directions, so that many potentially desirable architectures are evolutionarily unattainable, and many observed geometries reflect immutable underlying patterns of growth and development.

Gould explored these ideas in his remarkable studies of *Cerion,* the genus of land snails that I came to know in Curaçao and that also existed in a spectacular diversity of forms in the Bahamas. He interpreted the beehive shape of *Cerion,* with its slightly narrowed last whorl ending in an expanded adult outer lip, as the inevitable consequence of a complex three-phase pattern of growth. During the first phase, the shell expands in typical logarithmic spiral fashion, with each whorl larger than the preceding one. In the second phase, the shell's length increases, but the cross section of the whorls remain approximately constant, producing a cylindrical instead of a conical shape. Finally, as the snail approaches maturity, the whorl actually decreases slightly in diameter, and the adult lip then forms when spiral growth has ceased. Differences in shell shape arise from variations in the duration of each phase, but this variation in constrained within the limits of the unshakable underlying three-phased growth pattern.

In a formal architectural sense, this analysis is correct; but is it sufficient? What enforces the constraint? Is the prohibition against change in the pattern of growth required by a mechanical law or by mathematical necessity, or are departures unacceptable for ecological reasons? Gould's shells were removed from the ecological context of the snails that built them. But even in the field, where Gould has spent weeks or even months collecting them, *Cerion* yields unwillingly to observers eager to understand the world from their station. Sitting motionless on plants and rocks for months on end, they are as inert as any living creature in the wild can be. I can hardly fault Gould for celebrating *Cerion's* geometry in favor of the ecology to which it must be adapted.

My embrace of evolution as an adaptation-producing process does not imply that I interpret everything as adaptive. In all my writings on this subject, I have consistently and strongly argued that we need criteria to distinguish adaptations from incidental effects and phylogenetic vestiges. Adaptation should be neither uncritically accepted nor reflexively rejected. Like any other scientific concept, it deserves to be carefully examined with as many lines of evidence as can be mustered.

Of all the arguments against the importance of adaptation, the most potent to me was that the factors controlling the formation and disappearance of species act independently of those affecting the births and deaths of individuals. Did the everyday lives of individual organisms have any bearing on the sweep of evolutionary history?

I began to think about this question after reading Stanley's 1975 paper on species selection, but a coherent way to proceed did not dawn on me until a remarkable meeting that took place in Panama in June 1976. Egbert Leigh, who by then had moved to the Smithsonian Tropical Research Institute (STRI), invited five of us to his residence on Barro Colorado Island for a two-day discussion of the biology and history of the marine biotas on the Atlantic and Pacific coasts of tropical America.

The setting alone would have made this meeting one of the finest I have ever attended, but the participants made it even more so. Barro Colorado, a rain-forest preserve operated by STRI, became an island in Gatun Lake when the Chagres River was dammed as part of the construction of the Panama Canal. Our deliberations were punctuated by idyllic walks in the forest and by the Leighs' harpsichord, which both Edith and Bert occasionally played. Even during the discussions, the sounds of cicadas and the smells of lush unbridled nature drifted in through the open windows as if to remind us that the emerging ideas should apply to our understanding not just of the diversity of marine Panama but to controls on biological diversity as a whole.

Besides me, Bert had invited three paleontologists and an ecologist. Wendell P. Woodring and Axel A. Olsson, then both well up in their eighties but remarkably vigorous in body as well as in mind, trudged up the more than two hundred stone steps from the boat dock to the STRI compound on the island. With more than a century of research on the fossil molluscs of tropical America between them, they engaged in spirited arguments over the ages of important fossil-bearing formations and offered a riveting overview of the events their fossils chronicled. Alfred Fischer, erudite and incisive as ever, set the events of tropical America in a more global geological context and offered his interpretations of climatic and tectonic cycles that affected productivity and organic diversity worldwide.

The ecologist in attendance was Charles E. Birkeland. He had worked for several years on coral reefs in Panama and then joined the faculty at the marine lab in Guam in 1975. On his first day in Guam, Edith and I took him out for a quick look at the riches of the Pago Bay reef flat, but the grand introduction was spoiled when I was stung by a jellyfish. Birkeland combined a deep interest in comparative ecology with the discipline and striking originality of an experimenter. The work Birkeland described in Panama particularly caught my attention. By monitoring the organisms that settled as larvae on plates that he had set out underwater at several sites on the two sides of the isthmus, he showed that individuals grow in size much faster, and that grazing is more intense, on the Pacific side.

All the lines of evidence that came up for discussion at this extraordinarily productive conference—Woodring's and Olsson's patterns of extinction, Fischer's and Birkeland's arguments about the role of nutrients in reefs and other tropical ecosystems, and my preliminary information on predation-related architecture— began to come together in a coherent framework. In 1966, Woodring had already shown that the episodes of extinction accompanying or following the emergence of the Central American isthmus were especially severe in the Atlantic. Many species and genera of molluscs that occurred in both the Atlantic and Pacific parts of tropical America before the isthmus formed became restricted to the Pacific side as the result of extinction of Atlantic members. Only about one-eighth as many groups showed the opposite pattern of geographical restriction to the Atlantic. Did this pattern of extinction and survival, I wondered, reflect the availability of nutrients in seawater? Perhaps productivity declined in the Atlantic, placing many molluscs in jeopardy but enabling corals and other reef animals that contained oxygen-producing and food-making algal cells in their tissues to thrive. The eastern

Pacific, where nutrients remained abundant or even increased following the formation of the isthmus, would then have served as a refuge of many species that were unable to make their own food.

A paper published in the *Veliger* later that year by Edward J. Petuch of the University of Miami prompted me to look into these links further. On the coast of Venezuela, he had discovered a community of molluscs consisting of species close to or identical with forms otherwise known only as fossils. In other words, this area appeared to be like the eastern Pacific in providing a refuge for vulnerable species. Was it also productive?

Edith and I visited the Paraguaná Peninsula of Venezuela in July 1977. Few coasts match the desolation and discomfort of the area around Bahia Amuay near Punto Fijo. A hot wind blows hard out of the northeast, kicking up sand that stings every exposed inch of the body. That same wind, however, allows deep nutrient-rich waters to well up to the sea surface, creating conditions very similar to those in the Bay of Panama on the Pacific coast. Without this productive Venezuelan refuge, many more Atlantic species would have become extinct.

Although I treated the connection between nutrients and extinction in *Biogeography and Adaptation* in 1978, I began to pursue it systematically when Ed Petuch came to work with me as a postdoctoral fellow in 1980. Armed with an astounding knowledge of molluscs, the eye of an artist, and an unequaled collecting zeal, Petuch was a maverick. The hundreds of species he has described and the important historical and biogeographical conclusions he has drawn from his taxonomic work appeared to him almost as ready-made revelations in his intensely creative mind. Although he was far more often right than wrong, he lacked the patience to document trait by trait how his new species differed from other described and undescribed ones, nor did he back up his assertions with evenhanded arguments and systematically collected evidence. On his good days, when his mind raced through its vast storehouse of information, he was the kind of colleague and soul mate few of us get the privilege to know.

The paper I published with Petuch in 1986 in *Malacologia* confirmed and extended the hypothesized links among nutrients, extinctions, and predation-related architecture. Heavily armored molluscs in the Atlantic were more prone to extinction than the less well armored groups. Although speciation might also have contributed to the contrast between the armored Pacific and less heavily fortified Atlantic types, differential extinction seems to have played an important role.

In part through the remarkable efforts of Jeremy Jackson and his associates at STRI and of Peter Jung at the Naturhistorisch Museum in Basel, a great deal more is now known about the classification, fossil record, and physical history of tropical America; but the conclusions of our paper seem to have stood the test of the last decade. Our data and other data imply that processes affecting the death and survival of individuals also have repercussions for the rise and fall of species to which those individuals belong.

In the early 1980s, I decided to write a book on this subject. Books are, for me, in many ways much more satisfying to write than standard-format scientific papers, because they enable me to explore ideas broadly and deeply. I began reading hundreds of papers on subjects ranging from horns and combat in dinosaurs to the evolution of insect flight, gigantism in early land arthropods, the evolution of colony form in corals, the history of leaf and stem architecture in land plants, suspension-feeding in early echinoderms, the running speeds of cockroaches and centipedes, shell construction in barnacles and sea urchins, the habit of rolling the body up in a ball in trilobites and isopods, the ecology of freshwater planktonic crustaceans, biases in fossil preservation, the history of extinction, geographical and geological controls on climate and seal level, and much else. *Evolution and Escalation: An Ecological History of Life* was published by Princeton University Press in 1987.

As I wrote the book and presented its contents at seminars and meetings, I sensed keenly that the tide of scientific opinion was against me. On one side were ecologists who cared little about the evolutionary dimension of the field. On the other side were the evolutionists and paleontologists, many of whom were ready to jettison the adaptationist framework. I felt caught in a struggle of time.

As was true for the first book, the second showed how much still remained to be done. Many of the patterns I perceived required better documentation, and some of the inferences I drew about the adaptive significance and consequences of traits needed experimental confirmation. Important questions begged for answers. In what ways are extinction and speciation selective? How do wholesale invasions from elsewhere in the world affect opportunities for evolutionary change? What conditions set in motion the great episodes of biological diversification and innovation? There was more than ever to keep me busy.

Chapter 13

BARRIERS BREAKING DOWN

Complacency is the enemy of change. As we grow satisfied with our daily routines, slowly but surely we become hemmed in by barriers that we and others erect. The status quo may be comfortable, but it is also stultifying. A certain dose of novelty stimulates the mind and reinvigorates the soul. A change in place, exploring a new field, or the birth of a child made me notice things that I took for granted before. I might never have thought to ask why North Sea shells are chalky or why Dutch forests look manicured, if I had not had New Jersey to place the familiar surroundings of everyday into a different context.

On the home front, a fear of the unknown manifested itself most dramatically in the idea of raising a child. For years, the thought of fatherhood filled me with a sense of uneasiness and apprehension. Never having felt much attraction to children and certainly not to babies, I imagined years of bondage to an incomprehensible, altricial offspring.

The birth of Hermine Elizabeth on December 3, 1981 showed how absurd and needless my worries had been. From her first nasal cry just before dawn that day, I knew that a barrier had been breached.

I could play again without feeling self-conscious. Hermine and I built block towers, raced and crashed cars, read books, squirted water from a rubber clam, staged heterochronous battles between a Permian *Dimetrodon* and a Jurassic *Stegosaurus,* tracked a mechanical whale as it motored through the bathtub, and held contests to see who could roll plastic rings the farthest. I served as a convenient, mobile, and unpredictable climbing rack. Much later, Hermine and I conversed in our own language in which English words were pronounced and spelled backward.

Children tend to take the world as given. My blindness, though quite natural to Hermine, nonetheless posed an early and interesting problem to her intellect. When she was two or three years old, Hermine expressed great relief in having brown eyes. She could see; so could her mother, whose eyes were also brown, but I was blind, and my eyes were blue. Therefore, Hermine reasoned from the available evidence, blue eyes must confer blindness.

Hermine came on all our trips. In 1983, we spent the summer in New Haven, where I immersed myself in library work and writing at Yale. We rented an old rambling mansion with a lusciously overgrown garden full of wild flowers, pebbles, and daddy longlegs. At eighteen months, Hermine was grappling with the complexities of language. *Thistle* clearly referred to a spiny plant, one to be carefully avoided, but Hermine soon generalized the word *shishel* to any thing or action that caused pain. The word was spoken very earnestly whenever Hermine saw or felt something with the potential to hurt.

Taking cues from their parents, many children develop an early and lasting aversion to insects. One of the great natural wonders of the world ensured that Hermine was not among them. In early May 1987, the ground in our garden and around the neighborhood's trees took on an entirely unexpected appearance as thousands of small holes began to riddle the surface. An audible clue to the origin of the holes came on the sultry afternoon of the eighteenth. A distant, all-embracing, high-pitching whine provided a continuous and faintly disturbing backdrop to the bright song of robins, sparrows, cardinals, and mockingbirds. It was as if electric wires were vibrating at a high frequency. In the days following, the subtle whine grew in intensity to a deafening frenzy of individual calls. The culprits were three species of seventeen-year cicada, the adults of which emerge en masse for a few weeks every seventeen years to attract mates before they lay eggs and die. The larvae spend seventeen years underground, chewing on tree roots, totally hidden from view. The adult insects are so oblivious to danger and so utterly lacking in defense that it is a simple matter to catch them. Five-year-old Hermine quickly discovered to her enormous delight that the cicadas not only posed no danger whatever but would readily begin singing in her hand. Other singing insects—crickets, annual cicadas, and katydids—were infinitely harder to catch as well as more aggressive, and none would sing once caught. The experience taught Hermine that insects were just as interesting and lovable as the gerbil, rats, and rabbit that successively became the objects of her affection.

Hermine became a keen observer. In the summer of 1988 I took my family and Bettina to northern Japan, whose fauna of cold-water molluscs I was anxious to compare to that of the Aleutian Islands. While on a surf-swept rocky cape in southeastern Hokkaido, Hermine was the only one among the four of us to find a mussel, which she saw tightly wedged in a deep crevice. That same summer, she spotted a tiny pair of empty valves of the clam *Hiatella arctica* at the bottom of a deep tide pool in Maine.

The rest of domestic life was equally agreeable, so much so that I frequently worked at home. In the kitchen, the daily routine of preparing vegetables and potatoes or rice, which I often garnished with home-grown herbs, provided a welcome diversion from the day's academic pursuits. Even dish-washing and dish-drying—chores I would have avoided at all costs at Bussum—were a small price to pay for Edith's good cooking.

Despite my contentment, I felt stirring within me the urge to leave Maryland. My everyday state of mind is affected strongly by the quality of place, and I had grown increasingly tired of the frenetic ugliness around me. The setting of Guam's marine laboratory, overlooking the windward coast of the island and catching all the smells of land and sea on the northeast trades, daily gave me pleasure and satisfaction during the many months we spent on Guam. College Park, alas, did nothing of the kind. The Zoology-Psychology Building, the soulless fortress in which my little windowless office was hidden on the fourth floor, bore all the marks of a low-bid contract. Completed just before my arrival in 1971, it was about as charming as a sterile city block of apartments. Open doors protruded at all angles into the long hallways, and all the walls were lined with the smooth tiles I associate with bathrooms.

As so often happens in buildings with poorly engineered heating and cooling systems, flooding and leakage perennially endangered everything from rugs in the chairman's suite to laboratory equipment and books. Had my graduate student Loren Coen not been at the lab at six in the morning on one spring Saturday, water might have invaded my collections and destroyed dozens of Braille volumes instead of soaking only those items that lay within the two-inch tidal zone above the floor.

What remained of the lush countryside endured as tenuous oases among vast tracts of sprawl. Ugly cheap shopping malls and fast-food outlets linked by wide car-clogged roads covered hundreds of square miles. Nice sunny days were spoiled for me by the predictably high levels of ozone and by the pervasive odor of hydrocarbons.

To be sure, none of this is unique to the metropolitan Washington area, nor is it even particularly bad there. In fact, life in such places as southeast Florida, Manila, London, Rio de Janeiro, and Los Angeles would have been far more unpleasant. Nevertheless, I found my surroundings oppressive and degrading.

When Philip Signor called on Hermine's sixth birthday in December 1987, I was receptive to his inquiry. Would I like to apply for a position at the University of California, Davis? On my three previous visits to Davis, I had been struck by the comparative serenity of the campus and the adjoining town of tree-lined streets and small shops. The university had, in less than thirty years after its elevation from Berkeley's university farm to an independent institution, achieved world-class eminence in biology. Indeed, Davis could claim to be one of the very few academic institutions where molecular biology had not completely drowned out everything else of interest in the study of organisms. Most important to me, the geology department housed an active group in paleontology, a field that I alone represented at College Park.

Phil outlined a somewhat unusual situation. The Davis administration had created faculty positions somewhat crassly labeled "targets of opportunity" that were designed to be filled by exceptional scholars who could be easily dislodged from their current posts. As a first step, I would need to send a letter of interest, along with the usual academic biography. With this information in hand, the would-be departmental home would make the case to the university administration that my application should be allowed to proceed. Once permission was granted, the department would bring me there for the usual grueling interview. In the event the faculty liked me, they could vote to recommend the appointment, which a battery of committees and administrators would then have to approve.

The term "target of opportunity" troubled me. It evoked thoughts of favoritism toward the traditionally downtrodden—women, racial minorities, and the handicapped—and this disturbed me as much as did prejudicial rejection. Was this scheme an attempt to circumvent the usual standards in order to diversify the faculty simply to ensure representation of the minorities regardless of their particular merits and talents? Did such a program not imply that women and the members of other underrepresented groups were being judged separately because they could not compete with white whole-bodied men? Would this not create a second-class citizenry whose members would forever and inescapably be tarred with the unspoken allegation that they could not have reached their present status without the relaxation of norms? If so, then such well-meaning policies would serve only to reinforce and aggravate prevailing stereotypes while creating resentment elsewhere. The inequities of discrimination would be perpetuated, not redressed. If it is the goal of society to enable the disadvantaged to participate fully and on equal terms with everyone else in the whole range of occupations and life-styles, then fair standards must be applied consistently, and efforts must be made to see to it that people acquire the necessary tools through appropriate education.

The problem of unequal treatment had come up nine years earlier in a very different context. Late in 1978, the American Association for the Advancement of Science announced that it would hold a special session devoted to research by handicapped scientists at its general meeting in 1979. There would be a symposium reserved for those who were blind, deaf, confined to a wheelchair, or otherwise physically incapacitated. In a letter published in *Science,* I strongly objected. Why should the standards of participation in this most public of all scientific meetings be different for the handicapped than for other scientists? Was this not the "separate and unequal" doctrine that had so justifiably been rejected by the U.S. Supreme Court? The only way that handicapped scientists will be able to hold their own is if they can compete with nonhandicapped colleagues. If the work was deemed to be interesting and to deserve recognition at a meeting, why not have it presented alongside that of others regardless of personal particularities? Their achievements could then be a source of pride, not the spoils of a misguided guilt policy. Never again was such a special symposium convened.

Despite my reservation about the position at Davis, I decided to apply. An exhausting four-day interview, during which I spoke with members of three departments, followed in April 1988. By June, I had in hand a highly attractive offer from the department of geology, which I unhesitatingly accepted. Keeping my promise to teach at Maryland in the fall, I moved the family to Davis in December 1988, just in time to escape the worst of the east-coast winter.

The reality of the "targets of opportunity" program was considerably less troublesome than I had feared. As my application wound its way through administrative channels, I learned of at least one instance in which a potential "target" had been rejected on the basis of an unsatisfactory interview. A second, successful recruitment of a target of opportunity involved Peter Marler, a distinguished scholar whose work on bird songs is justly world-renowned. Marler belonged to no identifiable disadvantaged class. I was therefore satisfied that the university was subjecting potential candidates to the same high standards it applied to other candidates for faculty posts.

Besides, it was commonplace for such institutions as Harvard and Yale to woo individual scholars even in the absence of formal advertisements for an academic opening.

I felt intellectually rejuvenated at Davis. My sunny office, from which I could hear jays and white-crowned sparrows and even hummingbirds go about their business outside, was large enough to accommodate my ever-expanding collection and library as well as work space for examining specimens. Geologists and biologists with all sorts of expertise populated the campus. I could talk over plant evolution with James Doyle, brachiopods and phylogenetic reconstruction with Sandra Carlson, animal mechanics with Olaf Ellers and Richard Cowen, tectonics and the origin of continents with Eldredge Moores, the conditions surrounding the Cambrian explosion of life with Phil Signor and Jeffrey Mount, and insect biology with Hugh Dingle. I could consult Arthur Shapiro about butterflies and biogeography, S. Bradley Shaffer on salamanders and turtles, Marcel Rejmanek on plant invasions, James Quinn on theoretical biology, Rich Grosberg on invertebrate population biology and evolution, Thomas Schoener on island biogeography, Peter Moyle on fishes, Susan Harrison on conservation biology, Philip Ward on ants, and James Carey on insect aging and invasions.

There were improvements in daily life at home as well. Edith quickly created a lush garden, in which something bloomed every day of the year and from which I could harvest herbs whenever I pleased. In an effort to save some of the native biological communities nearby, she became active in protecting the Quail Ridge preserve, a hilly tract of dry woodland near Lake Barryessa in the foothills of the Coast Range. Hermine took up tap dancing and ballet and studied flute. The town itself was, to use a Dutch adjective, *gezellig*—its sense of community, created in part by the thriving twice-weekly farmers' market and by other events throughout the year, far exceeded that of any other place I have lived in the United States. Benefits rarely come without cost, and the move to Davis was no exception. Bettina would have to stay behind. I felt as if I were abandoning her, which after so many years of work seemed an especially heartless thing to do. Fortunately, Bettina's resourcefulness softened the blow, and she has gone on to a diverse and freewheeling career of writing and consulting.

It was important for me to find a suitable replacement for Bettina as soon as possible. The complex hiring procedure at the university would take a minimum of six weeks and might well drag on much longer if the right candidate did not come forward. I need not have worried. Of the thirty-eight applicants, several of whom had doctorates, Janice Cooper stood out as the obvious choice from the beginning. Like Bettina, she had studied English as an undergraduate, and she clearly had a gift for languages. After graduation from Duke University, she applied herself to science, geology in her case, and wound up as one of Richard Cowen's graduate students at Davis. Her diverse background was perfect for me. Sharp and inquisitive, Janice had strong interests in a host of geological subjects ranging from paleontology to volcanoes and earthquakes. Her facility with words and expressions was extraordinary, making her an informed and interesting commentator as well as a highly proficient reader. Besides English, she read expertly in French and German and managed very respectably in Dutch, Spanish, and even Latin. Her talent for organization transformed my office from a state of managed clutter to one of order and efficiency. With her second-to-none computer skills, Janice even dragged me partially into the modern age of word processing, electronic mail, and computer-aided bibliographic searches and calculations.

One of Janice's talents I valued most highly was her working knowledge of Russian. Two years before coming to Davis, I had become interested in the biological history of the cold northern oceans during the last fifty million years of Earth's history. I wanted to understand the consequences of the establishment of the Bering Strait between Alaska and Siberia some four million years ago. Very quickly and unsurprisingly, I discovered that there is a vast, impenetrable scientific literature in Russian on the living and fossil faunas of the Arctic and North Pacific oceans. Without this source of information, a study of the biology and history of cold northern seas would be hopelessly incomplete and unacceptably biased toward a western European and North American perspective. I knew no Russian, however, and very few of the relevant monographs and scientific papers had been translated into English. Janice turned out to be an expert translator of Russian into English and made dozens of important Russian works accessible to me.

The opening of the Bering Strait represents a good example of the disappearance of a geographical barrier to the spread of plant and animal species. Geographical barriers had played a guiding role in much of my research since the early 1970s. My interest in this subject began in the tropics. Today's tropical faunas differ in architecture as well as in the number of species. At least some of these differences came about because the regions in question had profoundly different histories. Today's four great shallow-water tropical marine provinces—the western Atlantic, the eastern Atlantic, the Indo-Pacific, and the eastern Pacific—are separated by barriers, either land masses such as the African continent or the Central American isthmus or deep ocean basins such as those of the Atlantic and Pacific. With few or no species able to cross these barriers, the four tropical marine regions developed more or less independently.

Barriers, however, are neither permanent nor totally effective. During warm climatic intervals of the last few million years, for example, many marine animals were able to spread around what is now the temperate coast of southern Africa from the Indian Ocean to the tropical Atlantic. Strong ocean currents enable the larvae of some animals to be carried great distances, so that some species occur on both sides of the Atlantic or in the western as well as the eastern Pacific Ocean. The land barrier in Central America has been in existence for only about three million years and might have acted to deter dispersal of some species since perhaps ten million years ago. Barriers come and go frequently. What effects do such geographical changes have on the world biota?

When species cross from one province to another, they encounter an entirely new array of potential friends and enemies as well as an unfamiliar climatic regime. They also change the conditions of life for the species in the recipient biota. Such changes potentially provide a wealth of evolutionary opportunities. If some types of species are better able to cross barriers than others, their lineages might have a greater global impact on evolution than do those species that stay put where they evolved. I therefore wanted to know how invading species differ architecturally and ecologically from noninvaders. Did the invaders owe their geographical imperialism to superiority in competition, defense, and fecundity, or were they drawn haphazardly as a random sample from the pool of potential invaders in the donor biota? Once established, how did the intruders affect the ecology and the evolutionary environment of native species? Did they cause extinction, or could invaders simply insinuate themselves into the gaps and crannies left relatively unexploited by the natives in the existing ecological architecture of the community?

Such questions had practical as well as academic significance. Commerce and human endeavors have been breaking
down geographical barriers on a global scale. Not only are
species being transported in soil, with plants, and in ships' bal-
last water, but canals have enabled species to invade from one
river system to another, or even from one ocean basin to another. With the completion of the Suez Canal in 1869, for example, hundreds of species from the Red Sea became established in the Mediterranean. What would happen if the
present-day freshwater Panama Canal were to be replaced by a
saltwater passage, which would link the Atlantic and Pacific
oceans for the first time in three million years? The apparent
biological sophistication of eastern Pacific as compared to
western Atlantic molluscs pointed to the hypothesis that invasion through a new seaway might be a predominantly one-way
affair toward the Atlantic.

Further insight into such large-scale invasion, or biotic interchange, would have to come from other parts of the world.
The northern oceans presented an especially attractive region in
which to pursue such studies. For tens of millions of years, the
drab cold-water faunas of the Pacific and the Atlantic had developed in their own world, geographically cut off from each
other by the Bering land bridge. The opening of the Bering
Strait between Siberia and Alaska about four million years ago
ushered in a period of intense interchange between these two
cold-water realms. The resulting pattern of invasion was strikingly lopsided; almost eight times as many species spread from
the Pacific into the Arctic and Atlantic basins than spread in the
opposite direction toward the Pacific. So dramatic was this invasion in the North Atlantic that almost all the common plants
and animals of the New England shore—mussels, limpets, periwinkles, dogwhelks, barnacles, sea urchins, sea stars, sand dollars, soft-shelled clams, eelgrass, and kelps—owed their origins

to the North Pacific. By contrast, only a handful of species from the North Atlantic had penetrated the rich communities of Siberia, Alaska, the American Northwest, and northern Japan. These patterns of invasion had been apparent to biogeographers for decades, but their ecological and architectural dimensions remained unexplored. I wanted to synthesize all available information on the taxonomy, distribution, architecture, evolutionary relationships, and fossil record of shell-bearing coldwater molluscs in cold-temperate and polar northern seas. The Aleutian voyage I described in Chapter 9 came during the early phases of this project; it was followed by trips to northern Japan in 1988 and Newfoundland in 1990, as well as by intensive work in museums, libraries, and my own collection.

At Davis, Janice began to translate relevant passages from dozens of Russian works, most of which had been ignored or overlooked by me and other western scientists. These publications revealed a very rich living fauna in the northwestern Pacific, as well as a fossil record that included many groups not known as fossils anywhere else. The political barrier that so long separated Soviet and Western scientists has fortunately broken down, greatly benefiting the joint effort that is so clearly needed to study the present and past biological diversity of the northern oceans.

The study of these cold faunas revealed an aspect of biotic interchange that I had failed to appreciate earlier; namely, the role of extinction. At Johns Hopkins, Steven Stanley had already shown that extinction affected Atlantic faunas much more dramatically during the last three million years or so than it did the Pacific ones. Even within the North Pacific, interesting and unrecognized differences in faunal disturbance existed. Nearly twenty common shallow-water molluscs living in the northwestern Pacific represent lineages that also occurred on American shores in the Pacific during Miocene and

Pliocene time, two to ten million years ago. Since the Pliocene, these lineages have become extinct on the American side. The northwestern Pacific therefore served as a refuge from extinction, much as did the tropical eastern Pacific and the Caribbean coast of northern South America. The northwestern Pacific contained the smallest numbers of foreign invaders, whereas the regions with the most dramatic impoverishment, notably the east coast of temperate North America, witnessed the influx of the largest contingent of intruders.

This same pattern—the most intensive invasion coupled with the most dramatic extinction—could be seen in all the other cases of biotic interchange on land and in the sea for which reasonably complete historical information was available. Competitiveness and defensive sophistication might indeed play a role, but the history of extinction explained all the known cases. The Suez invasions made sense in this light, because the eastern Mediterranean had been greatly impoverished by extinction during the last several million years, whereas the Red Sea probably was not. Invasions through a Panama seaway might still predominantly affect the Atlantic not only because of a difference in biological sophistication between Pacific and Atlantic biotas, but also because the Atlantic biota was more affected by extinction than was the Pacific biota.

Invaders in the sea typically did not bring about the extinction of native forms. Instead, extinction could trigger invasion. Here was an object lesson on the importance of history. Invisibility of a biota cannot be understood with reference to current conditions alone; it demands a knowledge of the past. The findings also carry a message of our own stewardship of the world's surviving species. The loss of diversity through human agency is bad enough by itself; the fact that it increases the vulnerability of surviving communities to invasion by foreign species, whose effects are sometimes quite dramatic, magnifies and greatly complicates this loss. I published these findings and ideas in a paper in *Science* in 1991.

If I smugly faulted ecologists for failing to appreciate history, my satisfaction could not hide the barriers that bounded my own intellectual enclave. Having been nurtured in the traditions of natural history and science, I thought about evolution in purely biological terms. Species interact with one another in communities and ecosystems, and the individuals comprising these species carry in their genes a complex set of instructions for building and expressing the phenotype. The replicability of the genes is influenced by the attributes of the phenotype and of the individual's environment. Implicit in this view is the expectation that we can achieve a complete understanding of organisms and their evolution once we know enough biology—life's molecular building blocks, the biochemical and developmental pathways by which molecules and morphology are constructed, the genetic basis of adaptation and species formation, the natural history of organisms, and the historical events of climate and tectonics that have buffeted populations and ecosystems over the long sweep of geological time. Our habit of expressing ideas in a vocabulary derived largely from biology predisposes us to think of evolution as a phenomenon unique to organisms. It obscures the enormous explanatory power and generality of evolution as a unifying concept in all branches of knowledge in which history is important.

Very gradually, I discovered that the language of economics offers a promising means by which these general features of evolution may be emphasized and explored further. Biological interactions—predation, competition, and cooperation—are played out among entities in a marketplace of resources. Such economic factors as supply and demand, risk and reward, cost and benefit, and trade influence pathways of adaptation. They affect the way in which essential resources are allocated to competing functions within the bodies of individual organisms, as well as among the organisms in an ecosystem. Beginning in the 1960s, Robert MacArthur and many other ecologists applied microeconomic allocation theory to the study of how animals search for food and of how energy is allocated to reproduction. Individuals have a certain amount of energy available to perform such functions as growth, reproduction, and defense. The amount of energy spent (or allocated), as well as the pathways taken to perform these functions, depends on local market forces—the availability of the resources, the benefits and costs associated with exploitation, and access to resources. For example, life histories characterized by early reproduction and large numbers of offspring may reflect a high per-capita risk of death for young individuals, whereas late reproduction and small numbers of offspring are linked with a lower per-capita risk of mortality. Patterns of energy allocation within individual organisms are akin to the everyday economic decisions made by people, for whom the balance between cost and benefit or between risk and reward is reflected in the price of a commodity.

Although this microeconomic approach adequately describes the behavior and adaptations of individual organisms, it offers little help in understanding how the large-scale economic conditions of supply and demand of resources in whole ecosystems affect evolution at the level of species formation, invasion, and extinction. Through their work on energy flow and nutrient cycling in ecosystems, ecologists like Eugene P. Odum and Howard T. Odum gained a keen appreciation for the macroeconomics of resources in nature, but these advances in the study of ecosystems went almost entirely unnoticed by evolutionary biologists, including me. Yet, slowly but surely, I began to grasp that there was a grand unity between the principles of evolution on the one hand, and the ideas of ecology and economics on the other. This unity is inescapable. Biological entities (cells, individuals, populations, and species) are economic entities that create, compete for, and modify resources. Economic principles are not merely metaphors for concepts in evolutionary biology; they are identical to them. Of course, there are differences between the economy of nature and the human economic system. Inheritance among organisms is chiefly by genes, whereas in human affairs it is cultural. Consequently, change proceeds much faster in the human economy than in the realm of all other organisms. Natural selection only operates in the present and cannot predict or plan for the future, whereas human decision-making can incorporate forward-looking thought. Still, the essentials—adaptation and selection, innovation and trade, and feedback between resources and consumers—operate in both the world of organisms and in the world of human commerce and civilization. As Darwin had shown long ago when he used Thomas Malthus' arguments about population growth being limited by food in his theory of natural selection, biologists can profit from the insights of economists. Equally appealing is the possibility that economists could derive lessons from an economic perspective on the history of life. We live in an age when economic health is held to be nearly synonymous with economic growth. Yet, economies cannot grow forever. Does the study of evolution and of ancient ecosystems offer any guidance about how to construct a healthy economy that is not expanding?

Had I taken the slightest interest in economics and not been so single-mindedly absorbed in science courses during my education, this link between evolution and economics might have become apparent to me long before. As it was, the preoccupation of economics with abstract models and with human cultural institutions such as corporations, stock markets, banks, government agencies, labor unions, the retail sector, and monopolies repelled me, and persuaded me that the subject was incapable of generalization beyond the human sphere. I finally perceived the connection between economics and organic evolution in the mid-1980s. It was the increasingly frenzied arms race between the United States and the Soviet Union that forced me to think about the broader implications of my work on biological escalation. The parallel between the evolutionary dynamics of snails and crabs and the military dynamics of the superpowers was too obvious to ignore. If I could try to understand how enemy-related adaptation starts and stops in the realm of organisms, perhaps similar ideas might apply to the development of ever more potent, ever more expensive machines of war. Perhaps the history of life could illuminate the mix of opportunities and constraints that controls budgetary and resource allocations to the development of weapons by governments. Some economists and historians, of course, had already been considering these questions from the limited perspective of their own disciplines, but they knew little about evolution, and had at their disposal only a very short historical record with which to test their ideas. With the broad sweep of time available to paleontologists who reconstruct the history of life, some processes and phenomena that are difficult to discern in a few thousand years of human civilization might emerge with greater clarity. At the same time, economists and historians can examine events and market forces on a global scale and with a temporal precision that is unattainable in most paleontological studies.

Better late than never, I began reading about markets, money, trade, corporations, and military history. As I delved into the so-called dismal science, I began to see that, just as economic stagnation is inimical to military escalation because money for arms and soldiers are perceived to interfere too much with other competing social demands, so adaptation and escalation between species and their enemies are prevented by adaptive gridlock and functional incompatibility between opposing selective forces. Economic growth makes military adventurism feasible. In our economy, such growth is fueled by trade, technological innovation, and a social climate of tolerance and permissiveness. In the history of life, the great bursts of reorganization and adaptation are spurred by analogous conditions of economic growth, in this instance involving increases in the availability of, and access to, energy. Greater availability and access lead to larger per capita energy budgets, which in turn allow absolutely larger outlays for one or more life functions such as locomotion, defense, feeding, reproduction, and the support of partnerships with other organisms in mutually beneficial arrangements. This kind of economic growth is, I believe, made possible by massive volcanic eruptions under the sea. By introducing gigantic amounts of the greenhouse gas carbon dioxide into the ocean and atmosphere without the climatic disruptions that often accompany eruptions on land, enormous volcanic outbursts under the sea warm the world's climate, expand the tropical zones to higher latitudes, and provide conditions favorable to the evolution of energy-intensive modes of life. The most spectacular episodes of undersea volcanism known thus far occurred during the early stages of the breakup of large continental land masses. The largest of these events took place some 130 million years ago during the early Cretaceous period of the Mesozoic era, when at least 51 million cubic kilometers of basalt rock formed as the result of unimaginably huge eruptions beneath the Indian and Pacific oceans over a period of two to three million years. This event, and other earlier and later episodes, coincides with known warm intervals in Earth's climate, as well as with times of biological innovations and species proliferation on land as well as in the sea.

Most satisfyingly to me, this perspective shows how economic principles make sense not only of natural selection and adaptation at the level of individual plants and animals, but also of evolutionary patterns at other levels of the biological hierarchy of organization, ranging from genes to species and beyond. Every biological entity is an economic entity; it lives, evolves, and dies in an economic context of resources, competition, opportunity, challenges, and constraints. The survival and reproduction of individual organisms are dictated by what might be termed microeconomic market forces of competition, predation, and partnership. The formation, spread, and disappearance of species and the more inclusive evolutionary units to which species belong can be understood in terms of such macroeconomic factors as the supply of, access to, and demand for energy and nutrients. These economic principles are inescapable; they imply adaptation. At every level of biological organization, entities must resist and respond to challenges ranging in frequency and severity from changes in enzyme concentration, varying by the minute or by the hour, to mass extinctions and tectonic upheavals that occur on the scale of millions to hundreds of millions of years. I published a summary of these economic ideas in *Paleobiology* in 1995.

Barriers, of course, remain. There are countless fields I shall never study, techniques I shall never learn, abilities I shall not have, and field sites and specimens I shall not see. Yet none of these barriers is unbreachable. I can collaborate with others who have access to the things that are out of my reach. As long as we recognize that barriers exist and think of ways to transcend them, we can expand our horizons and keep complacency at bay.

Chapter 14

INTEGRATION AND THE KNOWLEDGE ENTERPRISE

An account of the history of nations would reduce to a mere skeletal outline if it dwelled only on battles won and lost or on the succession of rulers and riots. In the same way, the story of a life would be hopelessly incomplete if it were told as a parade of actions and decisions. The full richness of history and of the course of a life derives from the circumstances and deeds of every day. It is this culture that so powerfully shapes our opinions, our outlook on the larger issues of the day, and our perceptions of how we fit into the world.

In this chapter and the next one, I want to describe my everyday work and to sketch how my experiences in research and teaching have influenced my outlook on important issues. I shall have things to say about education, the role of research at universities, the centrality of evolution as a world view, the health of scientific journals, my relationship to the blind community, and the value of museums.

How dull life would be if every day were like every other, if year after year I taught the same material in the same order, if my research were so predictable that I could hand an administrator a five-year plan and live by it. For me, the knowledge enterprise does not work this way. It is spiced with variation and surprise, enough to make me anticipate the next day's harvest as I would a good meal. I might happen upon an intriguing paper in a new journal, discover something new in a specimen I am putting away in my collection, receive a shipment of exciting material on loan from a museum, dissect an argument with a thoughtful student, or learn that a paper has been accepted for publication. More rarely, a new idea forces me to see old questions and observations in a new light and to add an item to my research agenda. For me, the everyday facets of scholarship—writing, editing, working with specimens, and talking with colleagues and students—merge with research in the field into a highly integrated, intensely creative endeavor. There is so much scope in science, so much freedom to explore and to work, that scientists can bring to their discipline the same individuality, the same personal signature, that novelists and painters bring to theirs.

Yet the public perception of science often contrasts sharply with this view. For many people, even for some insiders, science is impersonal; it is unapproachable, tedious, complex, difficult, spiritually degrading, and occasionally dangerous. The humanities civilize us; science is only a tool. The power and elegance of science disappear in a concatenation of jargon, statistics, sterile laboratory exercises, and received wisdom.

Were I a participant in "hard science"—research done by teams in well-run laboratories directed from the top by a few eminent scientists—my charitable views of science might crumble; but I work alone or with a few collaborators and students for whom I want the same freedom that my advisors entrusted to me.

Over the years, I have benefited enormously from a succession of more than a dozen superior doctoral students. By working with me yet independently from me, they greatly expanded my own intellectual horizons and encouraged me to think about new problems, new techniques, and different groups of organisms. Their dissertations dealt with all sorts of topics: predation of snails in the Potomac River, the ecology of large whelks in Florida, the evolution and classification of gall crabs and their coral hosts on tropical reefs, the role of seaweed-eating spider crabs in the ecology of Caribbean reefs, communities of animals living as borers in snail and hermit-crab shells in Guam, the evolution of sharks, the muscle physiology of crab claws, communities of animals living on the shells of brachiopods hundreds of millions of years ago, and more.

I inherited from my Yale advisors the belief that independence of mind lies at the core of a successful scientific career. Students who came to me could expect advice, references to the literature, and encouragement, but not day-to-day management. I insisted that they alone should choose the topic of their dissertations. After all, the dissertation effort tests every ounce of a student's resolve, and therefore it had better be deeply interesting to its author. This hands-off attitude works well with highly motivated people who want to make their way in the research world, but not with some beginners or with those who have grown too accustomed to the spoon-feeding syndrome that infects so much of education at every level today. I always interview prospective candidates to evaluate their motivation and scientific curiosity.

I looked to my Princeton experience for hints about teaching undergraduates. The professors I most admired presented their fields as dynamic disciplines, rich with the personalities of practitioners. Their lectures brimmed with unanswered questions, promising directions for future work, and above all with their own often very substantial scholarly achievements. Like them, I wanted to demystify science, to make it accessible, to penetrate the jargon and the formulaic style of its literature. There was no point in repeating what could be better read in a textbook. Instead, I endeavored as my Princeton mentors did to present and connect facts and ideas in a form not readily available in published sources.

Occasionally, these attempts led to unexpected outcomes. One spring day in 1987, while lecturing on the evolution of vertebrates, I observed parenthetically that leaf-eating birds are overwhelmingly flightless or weak fliers. I had never thought about the possible explanations for this apparent fact, but I deemed the matter worth a brief digression. On the podium, related observations went through my head and out of my mouth. Like most dinosaurs, the group from which birds most likely descended, leaf-eating birds break down fibrous plant tissues with grit and stones in the gizzard rather than with grinding teeth in the mouth, as most mammals do. Heavy equipment for chewing and digesting plants that are high in bulk but low in nutrients is incompatible with the demands for a light build in a flying animal. Moreover, rates of metabolism in leaf-eating birds are low, as they are in tree-dwelling mammals such as sloths. Did functional conflicts among energy demands, weight restrictions, and powered flight preclude a leaf-eating habit in flying animals? The ensuing discussion, lively and wholly unplanned, lifted the audience out of its usual passive state. The question remained unanswered, but the act of asking it communicated an important lesson. There is much we do not know, and often we can imagine solutions that even beginners can appreciate and tackle. Science need not always be beyond reach. We can beat back ignorance even if we are novices, if we ask the right questions and apply our minds to gathering the evidence that is required. Later, I took up the matter with Robert Dudley, a leading authority on the biomechanics of insect flight, who also happens to be Bettina Dudley's elder son. We coauthored a brief paper in *Functional Ecology* in 1992 in which we showed that the power requirements for active flight are indeed incompatible with the heavy weight of foliage and the digestive system required to process it. We also confirmed the impression that animals that swallow leaves for food are poor fliers.

For many years there has been a steady chorus of criticism that universities emphasize research at the expense of instruction. Scholars in all disciplines, so the argument goes, are so engaged in raising funds and in publishing papers that they have little time for students or for preparing lectures. There is, of course, an element of truth in such a portrayal. As the number of scientists swells and the available research funds are increasingly eaten up by huge bureaucracies at the universities and the funding agencies, competition for the diminished resources is intensifying, and appallingly large blocks of time are spent writing and reviewing lengthy grant proposals in which every methodological detail and every expected result must be carefully spelled out. Tenure and promotions cannot be had without publishing papers, an activity that demands time and concentration. Yet I believe that research is an essential ingredient in the best university teaching. There are gifted teachers who are inactive in research, as well as accomplished researchers who are dreadful teachers; but the best teachers are those for whom inquiry and communication are inextricably linked. Calls for increased teaching loads and for correspondingly reduced scholarly activity are off the mark. They reflect a profound misconception of teaching as a one-sided flow of predigested information from an unchanging corpus of doctrine developed by a master to a passive student. If critical thinking and the weighing of evidence define science, and indeed almost all other academic disciplines, we as teachers must be deeply engaged in the process of learning, and we must preach what we practice.

Although I experience lecturing as a satisfying mix of synthesis and oratory and delight in stimulating those who are already committed to the pursuit of knowledge for its own sake, other aspects of teaching have come less easily. Many students arrived with so little motivation and ambition that I felt powerless to reach them. Other, far more gifted teachers such as Richard Cowen at Davis prod, cajole, and awaken the unwilling, and succeed in spite of a deeply flawed system. I felt defeated by society, which tolerates large impersonal classes, prescribes low and inconsistently applied standards, pits athletics against academics in institutions of learning, and often measures the reputation of universities by the prowess of their sports teams rather than by the quality of the degrees those institutions confer on undergraduates.

As the great unifying theme of biology, evolution has permeated all my teaching. From the design of antibiotics and the epidemiology of viral diseases to the development of language and the structure of our social system, everything about our species and about other creatures makes sense in the light of evolution. As long as heritable entities vary in survival and propagation by virtue of traits that undergo trial-and-error selection, adapted systems capable of resisting, responding to, and ultimately predicting and modifying their environment will arise. We as humans are unique in many important ways—we have a greater impact on our surroundings, and we can forecast and alter our circumstances as no species before us could—but we are the product of natural selection, and we cannot escape evolutionary realities. Even the freedom to engage in scientific inquiry or to deny our phylogenetic link with the rest of the animal kingdom springs not from some mysterious irreducible force but from the modification and elaboration of traits inherited from our ancestors. I believe that no educated person can remain ignorant of evolution and its all-encompassing implications.

Students have not always shared my enthusiasm for the primacy of evolution. During the 1970s and early 1980s, proponents of what was misleadingly called creation science, or scientific creationism, were waging a loud campaign to have the Biblical stories of creation sanctified as science in the public schools. Each year, several students came to me proclaiming that they should not be held responsible for any evolution-inspired science I might teach, nor should they be forced to apply it on examinations. I replied by saying that, just as I must endeavor to understand the religious position of those who do not accept evolution, religious creationists must understand the scientific principles of evolution if any meaningful discussion of the subject is to take place.

The enormous popular appeal of scientific creationism was brought home to me when several of my colleagues and I agreed to take part in a public debate at Maryland in 1975 with one of the leading zealots of that movement. I choose not to reveal his name in order to perpetuate the obscurity this man so richly deserves. In front of an audience of twelve hundred people, most of them creationist partisans, our adversary trotted out the usual arguments. The six-thousand-year-old Earth, he said, had not existed long enough for the slow process of evolution to produce the known diversity of life. Somehow, he dismissed out of hand the enormous weight of evidence in favor of the Earth's great antiquity—perhaps five billion years—and for the very early appearance of life on our planet some three and a half billion years ago.

Evolution violates the second law of thermodynamics, our adversary continued. According to this law, order disintegrates into chaos; yet evolutionary theory holds that order increases. The ingredient our adversary missed is that energy moves throughout the living system. The second law of thermodynamics applies in systems from which energy is lost, but in living creatures a steady stream of energy is pumped through the system. When energy flow stops, cells die and turn to dust or, in thermodynamic terms, to chaos.

The third, and potentially most powerful, argument was an appeal to credibility. How could chance alone produce such complexity as we see in the vertebrate eye, the human brain, the complicated life cycle of a parasitic fungus, or the social structure of ant colonies? Complexity may seem improbable without the guiding hand of a creator, but computer simulations show how easy it is to create remarkably complex and highly adapted entities from a program that "learns" from its mistakes through a process of trial and error. "Blind chance," as it is often gratuitously labeled, is not really "blind" at all. Variations that arise through mutation are filtered through the organism's established pattern of development and through the environment, so that only those that provide a survival or reproductive benefit persist in the next round of selection.

As I should have expected, the debate turned less on scientific issues than on the source and credibility of knowledge. Should we believe the word of God in the Bible or merely the scribblings of mortal scientists?

When in subsequent years I received handwritten invitations to attend presentations by scientific creationists, I declined. The presence of scientists at such gatherings legitimizes the creationist cause, for it suggests as nothing else can that creationism is an important idea worth being considered by Biblical literalists and evolutionary biologists. By ignoring creationists on the turf of their choosing, we rob them of credibility. Scientific creationism and its variations must indeed be fought by scientists, but on our own terms and not under rules dictated by our enemies.

One such opportunity came in 1982. A bill that would require the teaching of Biblical creation as science in the public schools was introduced in Maryland's House of Delegates. Along with Steven Stanley from Johns Hopkins, University of Maryland historian of science Steven Brush, and many others, I testified vigorously against the bill. I had no objections whatever to creation as long as it was part of a religion that I was free to believe or not, nor did I mind reciting the creation story by singing Haydn's Creation Mass; but it isn't science, and no law will make it so. Fortunately, the bill failed.

I understand well enough why Christians and many others outside and even inside science would fear evolution. It clashes irreconcilably with a belief in a supernatural deity who directs, or at least sets into motion, all causalities of the universe. Such a god represents top-down control. Evolutionary thought perceives the regularities and complexities of life as the products of the repeated application of trial-and-error algorithms that by themselves have no higher meaning but that can produce creatures who attach purpose and significance to their lives. One might elect to interpret the various accounts of Genesis loosely, but how do we know which passages to take literally and which to revise?

It is equally clear why religion has been such a powerful force in human affairs and why its proponents so fiercely defend it. The morals and teachings of religion bind individuals to a social contract and enable society to function not merely as a collection of self-interested individuals, but as a culturally unified body. Even greater social cohesion is achieved when the moral code is seen to emanate from superhuman authority whose instructions must be obeyed without question. Societies whose systems of belief engender allegiance and obedience most willingly may be the most successful entities in the political and economic arena. In this sense religion and morality may be understood as attributes of an intelligent, social species in which the power of individuals to predict and modify the environment is amplified enormously by the actions of cohesive, organized groups. In a highly competitive world, religion succeeds as an effective means of group competition by providing a decided advantage not only to the society in which it is practiced, but also to the individuals who belong to that society.

To me, the evolutionary view of life does not strip away purpose and meaning, nor is it in any way demoralizing. I marvel at the complexity and beauty that can spring from simple rules, and I admire science as a way of describing and understanding that richness. The desire to contribute to a common good, to make society work, to champion a cause, or to celebrate the grandeur of the universe is no less worthy when it arises as an emergent property from molecules or atoms than when it is slipped to us by the creator.

It is a pity that those committed to a belief in the supernatural so often feel compelled to justify their doctrines as science even as they reject science as a way of knowing. For me, the explanatory function of science and the moral function of religion are complementary. Where science proposes and tests explanations with evidence derived from observation and experiment, religion should temper individual wants with the ethic of the common good and the principle that we treat others as we would be treated ourselves. Too often, religious and pseudoscience purport to offer explanations derived not from observables but revealed by an unassailable higher authority or force in whose absolute power and infinite wisdom we are compelled to believe. As a scientist, I reject such untestable doctrine. Dogmas that resist criticism and modification breed an attitude of intolerable, insufferable superiority. By embracing such a world view, individuals and society relinquish responsibility for their actions to an unknowable, unaccountable god or, worse yet, to a human dictator. How many lives have been lost, how many cultures trampled, how many landscapes defaced, in wars whose economic causes were amplified and corrupted by the fervor of those who were convinced that they alone possessed absolute truth or claimed a divine being on their side? Religion has a place, but we must harness it. Can we not express and enforce a body of morals such as the Ten Commandments, and achieve social cohesion and the common good, through democratically controlled government alone, without appealing to the beguiling yet destructive doctrine of absolute truth invested in a supernatural being?

Teachers and writers cannot expect to communicate synthetic subjects such as evolution effectively to an audience unless they thoroughly grasp their own message. Some scholars achieve such comprehension in a flash, but I am not one of them. I need time to read, to reflect, to ponder idly. Too often, however, the luxury of time falls victim to a frenetic pace and to escalating demands that we do everything superlatively. It has become fashionable to be seen rushing from one crucial meeting to another, reviewing dozens of papers and proposals, juggling research commitments with committee obligations, governing the department and university, and supervising a large contingent of students. In this environment, I have learned to say no frequently and firmly. My judgments of what matters and what does not might not be made by others, but one must make choices about how to use limited time most effectively in the context of one's own talents. For me, time to reflect makes everything else possible, and I safeguard it jealously.

I devote hours each day to reading the scientific literature. Knowing what other scholars have accomplished allows me to place my own inevitably limited contributions in context and enables me to inform my students of new developments. Is a shell I cannot identify really an undescribed species, or have I simply overlooked the obscure monograph in which it was named fifty years ago? How do my observations mesh with those of others? Does the observation I offered in my paleontology course last week for the post-Paleozoic decline of brachiopods still hold in the light of newly published evidence? The curiousity that drives my own research also compels me to read.

The vastness of the scientific literature is so overwhelming that scholars are constrained to specialize in one or only a few narrowly circumscribed fields. To varying degrees, we accept defeat in all the others. For the sighted reader, the task of acquiring expertise is facilitated by the printed word; only time and tenacity stand in the way of gleaning knowledge from the hundreds of thousands of journals, books, and databases that are accumulating in the world's libraries. Almost none of this information, however, is in Braille, nor should it be. Braille is bulky and expensive to produce, and all scholars know that most publications contain a great deal that can be profitably ignored. One obvious solution would be to buy one of the increasingly sophisticated machines that translate printed text into spoken words or Braille, but there are many problems. Reading with such machines is slow, and the translation is rarely accurate. A machine has difficulty with technical terms and mathematical symbols as well as with foreign languages. Most important, it is not versatile. Reading is limited to where the machine happens to be located, and the machine is not effective at scanning text for particular items of information.

As in much else, I have opted for a low-tech solution. No machine can yet match the accuracy, speed, knowledge, versatility, and pleasingly varied inflection of a competent human reader. I can take extensive notes on the Perkins Brailler, or even the slate and stylus, while someone reads to me. I can ask the reader to look for something particular—an entry in a long bibliography, a number in a complex table, the name of a species—and I can skip, repeat, have a word spelled, and ask for clarification whenever necessary. Mathematical symbols, unusual fonts, foreign languages, and tables and illustrations are easily read or interpreted by a knowledgeable reader. I find it far easier to take down information from a reader than from a machine or a tape recording. And not least, I can laugh and groan with a reader as I encounter something silly or pretentious. Imagine doing that alone, in the presence only of a machine utterly oblivious to the meaning of the words its artificial voice utters.

Edith, Bettina, and Janice are all fast, accurate, knowledgeable readers who bring beautiful diction and a lively intonation to their skill even under the most trying of circumstances. They deal with a barrage of improbable names, complex equations, and copious jargon. Authors who coin such species names as *Tumidimytilus chejsleveemensis* (a Miocene mussel from Kamchatka), *Lentigodentalium pseudomutabile* (an early Miocene tusk-shell from Germany), or *Nipponopanacca brevicaudata* (a deep-water clam from Japan) probably never imagined that their dissonant creations would ever be spoken out loud.

The result of this reading is a Braille library comprising notes on more than twelve thousand papers and books gathered together in more than two hundred thick volumes. Sometimes I take down only the summary of a paper, but more often I transcribe the bulk of the data as well as details of the arguments and interpretations. I retrieve material with the aid of an extensive card file by author, along with an elaborate cross-referencing system categorized according to subject, habitat, geographical region, and taxonomic group. To complement this library, I maintain a growing collection of print publications, and I make it a habit to scan the library shelves or their computerized equivalents for the latest books and periodicals.

I write manuscripts the old-fashioned way. I begin with drafts, usually dozens of them, composed with the Perkins Brailler on thin standard-sized paper scavenged from old manuscripts and memoranda. When more or less satisfied with the text, I prepare a version on an ink electric typewriter. Before the manuscript is sent off for review, it is thoroughly proofread and corrected, and I usually make further changes. At Davis, Janice Cooper has begun to scan typed manuscripts into the computer, so that large and small changes are easily incorporated.

To those who are at ease with computers, this method must appear a primitive throwback to the outmoded ways of an era best forgotten, but it works for me. The limiting step in writing is not revision or editing, but finding something to say and then putting words to paper. Revising my prose again and again enables me to polish and to craft a piece that says what I wish it to say.

Seeing how my own manuscripts benefited greatly from passing the gauntlet of reviewers and editors, I concluded that involvement in the publication of scientific journals held promise as a potentially rewarding way to serve my profession. I began with *Paleobiology,* one of two periodicals published by the Paleontological Society. Thomas Schopf and Ralph Johnson at the University of Chicago cofounded it in 1975 as a vehicle for synthetic and analytical papers on the fossil record. After a dozen years or so at Chicago, the journal moved to Davis, where coeditors Richard Cowen and Phil Signor were shepherding some seventy-five papers per year through review and publication. I joined as coeditor in 1990 and shortly thereafter took over full responsibility as editor. Teresa Carson, a highly competent and efficient woman with a dozen years of editorial experience with the *American Naturalist,* joined me as managing editor. While I concentrated on finding reviewers, evaluating manuscripts and their reviews, and communicating good and bad news to authors, she copyedited, modified journal format, and dealt with printers.

When my term expired in 1992, I became editor of *Evolution,* and Teresa transferred with me. This journal, founded in 1946 by the Society for the Study of Evolution, receives roughly four hundred fifty manuscripts per year on all aspects of evolution. These are handled by some twenty to thirty associate editors. These scientists, who like the editor volunteer their services without pay, are leaders in their respective fields and carefully read the papers assigned to them. They must also solicit and evaluate reviews of manuscripts by still other experts. During my three-year term as editor, I enlarged the journal and tried to attract thoughtful papers that would broaden the appeal and highlight the synthetic nature of the field.

Most people recoil at the thought of editing a journal. Too much work, they contend; too much time, too much disappointment to dispense. This is indeed not the job for the disorganized or for those who anguish over decisions. By delegating most of the critical day-to-day running of the journal to the managing editor, I typically spent not more than an hour per day reading and writing letters, evaluating reviews, looking over manuscripts, and handling a few papers myself.

I have suffered my share of rejections, and I appreciate the importance of diplomacy and kindness in dealing with authors. Constructive criticism of one's prose, data, and interpretations is bruising enough; gratuitous attacks, insults, and niggling editorial corrections covering the page discourage and hurt people and often cause distraught authors to give up and throw away salvageable work. The aim of the whole tortuous enterprise is to improve manuscripts, to make them readable, to send into the world contributions that beckon the journal's subscribers to read on. Even the manuscripts that are not accepted, which account for 60 to 70 percent of those submitted, usually contain much of value, even if they appeal to a more specialized audience. A few disappointed authors took great offense at my recommendation that they submit their revised papers to specialty journals. How dare I think that their paper should be sent to a second-rate periodical if it did not belong in *Evolution*? I found such responses and the underlying premise disappointing. Most scientific journals subject manuscripts to review. The main difference among them is not in the quality or credibility of the published articles, but in the size and the diversity of the audience. Important papers appear in all sorts of journals, including very obscure ones. Their quality should be judged on content rather than on the title of the periodical or book in which they appear.

In a system in which scientific papers are so carefully scrutinized before publication, there is always the danger that new, radical, or controversial ideas will be rejected prematurely by overzealous gatekeepers. As editor, I was keenly aware of the conservatism inherent in the review of manuscripts. I therefore occasionally accepted papers that conflicted seriously with established dogma even if these papers did not always meet all the usual standards of scientific rigor. As long as authors marshaled evidence and plausible arguments and refrained from making unsupported assertions, I felt that new ideas deserved a fair hearing in published form by the scientific community. If scientific rigor were the criterion prevailing over all others, many promising avenues of inquiry might not be pursued, and authors would quickly learn to publish only their safest and dullest work.

Academic conservatism is even more pervasive in the writing of grant proposals. Authors whose offbeat papers are rejected by one journal can often get their work published by another if they are diligent and willing to modify their prose in accordance with reviewers' advice, but the writers of controversial or innovative grant proposals usually face rejection without recourse or appeal. As per capita funding for science has declined, competition for the diminished resources by a greater number of applicants has driven the funding agencies to demand so many preliminary results and methodological details for a given proposed project that only a fool would propose exploratory research or a radically new line of study. It is far safer to propose work that builds on what has already been done and whose probability of successful completion is high. To me, the most interesting science does not proceed in this way. Research often takes unexpected directions, some of which turn out to be far more fruitful than the project that was originally conceived. Scientists should be encouraged to proceed in new directions even if success cannot be guaranteed and results cannot be predicted. But just as competition among organisms in the face of limited resources leads to adaptive gridlock, so the severe competition that characterizes science today may enforce a conservatism and a bias toward the predictable and the conventional that flies in the face of the creativity that is so central to the conduct of science.

Editing meshed well with my disinclination toward group participation. As one who remains ill at ease in large gatherings and who chafes at deciding scientific issues through compromise and consensus, I prefer to speak out and to act as an individual. I see this as a reality rather than as a virtue. Collective action, especially by scientists, has the potential to inform and influence society as we grapple with environmental destruction, human overpopulation, depletion of resources, military spending, and the preservation of cultural and biological diversity.

I feel a special responsibility toward my fellow blind, but in this arena too I have chosen individual action over large-scale group participation. Given my views on blindness as a nuisance rather than an insurmountable tragedy, I quite naturally gravitated toward the philosophical position of the National Federation for the Blind (NFB). This civil rights organization, founded in 1940 by Jacobus TenBroek and his followers, has diligently and effectively fought for the rights of the blind throughout its distinguished history. It has mounted a remarkably successful campaign to educate both the blind and the sighted about Braille, independent travel, and the limitless possibilities in employment and social integration as long as the blind are given the requisite tools, techniques, and positive attitudes. The NFB promotes Braille literacy, confronts government and private agencies when they treat their blind clients as inferior wards, opposes the creation of phony standards by accrediting bodies that evaluate schools and agencies for the blind, pressures sheltered workshops to pay their blind workers decent wages, fights cases of discrimination in court, crafts legislation, publishes the influential monthly *Braille Monitor*, and sells by far the best white canes on the market. A succession of brilliant leaders—TenBroek, Kenneth Jernigan, and Marc Maurer among them—has guided the NFB with a deeply sound philosophy, expressed with force and unusual eloquence in a host of speeches and articles. With an effective hierarchical organization of chapters in affiliates that cover the land from coast to coast, the NFB has become a powerful force in the affairs of the blind. It has successfully organized blind workers into labor unions, sponsored conferences on emerging technologies, created an international center for the evaluation of such technology, and stimulated education in many ways, including the creation of hundreds of scholarships. The NFB has even opposed features designed by others for the purported safety of the blind. Audible traffic signals—beeps, buzzes, and whistles activated as traffic lights change—were meant to inform blind pedestrians of when to cross streets and when not to venture out, but the NFB rightly points out that the signals interfere acoustically with the much more informative and reliable sound of traffic and are therefore confusing and dangerously misleading.

I support the NFB financially, encourage others to join, and endorse all its positions; yet I have contributed little to the movement itself. Now and then, I write articles for the *Braille Monitor*, give speeches, attend conferences on education and employment, talk about blindness to schools, and willingly offer advice and encouragement to anyone who calls or writes. The largest audience to which I have ever spoken was the NFB convention in Chicago in 1988. Perhaps selfishly, I hope that, by making my way as a scholar in the world of academic science, I contribute to a climate in which blind people have as much choice in life as do the sighted.

Just as teaching and research unite scholarship into an indivisible enterprise, and the extraordinary diversity of living beings and of biological phenomena reveals a deeply satisfying unity when set in an integrated evolutionary framework, so one's personal characteristics and accomplishments blend into a single integrated life, and individuals of diverse backgrounds integrate into a single, if complex, society. As scientists, we study the parts of things as well as the properties of the structures they constitute. As a society, we value individual qualities and benefit from the freedom to apply those qualities to the common good.

The integration of disadvantaged groups into society will have been achieved only when a person's accomplishments are judged for what they are, independently of that individual's characteristics. I am not ashamed of my blindness, but neither do I wish all my contributions to be identified with it. Blindness is but one of my attributes, along with being stubborn, Dutch, male, scientifically inclined, and a host of others. If I am to be a role model, to use a bit of current jargon, I want it to be for integration.

Chapter 15

CURATOR

Shells have always been more to me than just beautiful variations on an elegant theme of spiral architecture. These works of art, and the environments in which they were fashioned, have shown me the way to some of the larger questions in biology, questions about function and evolution and construction. Almost all my research began with observations in the field and on specimens in my collection. Shells hold as many surprises and prompt as many questions now as they did when I filled my first cigar box with them in 1957.

One winter morning in 1995, I brought some specimens of *Donax carinatus,* a sleek smooth beach clam from the Pacific coast of Panama, to demonstrate some points about function to my class. When the clam lies buried in sand, it is just beneath the sand surface, and its long streamlined front end points down. In this species, a sharp keel (or carina, hence the specific name) sets off the horizontal posterior or rear surface from the vertical sides of the shell's anterior, or front end. Work by Steven Stanley on other clams with a similar keel had

shown that the broad posterior surface stabilizes the clam in
sands that often shift as waves break or when animals burrow
nearby. The valve edges are decorated with evenly spaced
toothlike serrations, which form a tight seal when the valves
are shut. As I was idly fingering the specimens while waiting
for the students to settle into their seats, I observed a subtle
feature of these serrations that had previously escaped my notice. Whereas the serrations along most of the margin were
symmetrical, their flanks descending at the same angle on either side of each serration, those on the posterior edge were
strongly asymmetrical. There, the flanks on the underside of
each serration sloped very steeply, whereas the back flank fell
away gently from the crest. After class, I checked other species
of *Donax*. Some, such as *Donax lenticulatus* from Jamaica,
showed the same pattern; but *Donax vittatus* from Wales displayed symmetrical serrations throughout. I had absolutely no
idea what, if anything, this observation meant. Was the asymmetry of the serrations on the back end in some way adaptive,
or did the pattern merely reflect something about the construction of the tiny radial ridges that produced the serrations?
Could the pattern of serration be used to identify taxonomic
groups or to infer evolutionary relationships within *Donax*? I
did not know, but I wrote down the observation for possible
use in the future.

Something very similar happened a few months later while
I was engaged in conversation with a visitor. As we were ad-
miring a collection of Pliocene bivalves from the Pinecrest beds
of southern Florida, I picked up a specimen of *Andara rustica*,
a striking species immediately recognizable by the thick, square
ribs running radially along the shell from the valve beaks to
the edges. Suddenly, I noticed that the concentric lines that
cross these ribs at regular intervals were asymmetrical. The
edges facing the shell margin were sharp, whereas the edges

facing the hinge, opposite the free margin of the valve, sloped obliquely. When I followed one of the square ribs from near the hinge to its ending on the valve's margin with the tip of my finger, the succession of concentric riblets felt smooth; tracing the rib in the opposite direction, my finger encountered resistance, much as it would if I stroked a dog from back to front or if I ran it along a blade of grass from tip to base. This particular kind of asymmetry is, I believe, quite unusual in bivalves. Most other clams in which sculptural elements parallel to the free margin are asymmetrical have the dorsal slope, the one toward the hinge, steeper and therefore rougher to the stroke of the finger than the ventral edge. That kind of asymmetry is quite common and, as Steve Stanley was the first to show, prevents back slippage when clams take burrowing steps as they dig into sand. I do not know what the reversed asymmetry of the concentric sculpture in *Anadara rustica* means, nor do I know of other fossil and living bivalves that might have this characteristic. In retrospect, the attribute is quite obvious, yet neither I nor anyone else has ever commented on it.

I cannot claim to observe shells better than others do, nor would I pretend to discriminate more easily among species on the basis of shell features than other malacologists do. Every observer brings to his or her own science a unique perspective, and I am no exception. Inspection of an object with fingertips, fingernails, and thin needles reveals not only the broad outlines but also the small-scale details—the number, relative size, and orientation of elements of sculpture; the placement of teeth and folds surrounding snail-shell apertures; the pattern and asymmetry of clam-valve serrations; and the like—that a causal observer is apt to overlook.

Observation by hand is particularly well suited to objects the size of most shells. Although the tactile sense is fully adequate to capture the overall form as well as the subtleties of all but the tiniest shells, it is inadequate for obtaining an instantaneous impression of very large objects. The hands integrate information on a small spatial scale, whereas the eyes can take in objects over a very large range of sizes. Casual acquaintances are surprised when I fail to respond with enthusiasm to large works of sculpture. Most sculptors work to please the eye, and only when they adorn their creations with small-scale relief on a smooth fine-grained surface such as marble or polished wood do they appeal to my tactile esthetic. The fact that such pieces are stationary further complicates inspection, for it is the ability to turn shells in the hand that allows me to appreciate their form and relief from many perspectives.

Careful observation demands discipline and time. I must consult a specimen again and again before I have wrung all the information out of it. The acts of cleaning, measuring, and putting away specimens provide opportunities to get to know them better. By the same token, playing with data exposes errors and inconsistencies that might go unnoticed if the numbers were mindlessly and uncritically fed into a machine.

The overriding benefit of having my own collection is that I can work independently. Each of the tens of thousands of lots bears a Braille label of two or three lines, on which the species name, data and location of collection, and habitat are indicated. Thick Braille field notebooks record further observations on habits, diet, predators, co-occurring species, and other details.

My collection and library cannot, of course, match the holdings of a large museum. With the University of Maryland's College Park campus located only nine miles from the Mall in Washington, I took full advantage of the incomparable resources of the Smithsonian's National Museum of Natural History, the greatest of its kind in the world. This majestic place, housed in a massive building on Constitution Avenue, surely represents one of the great cumulative intellectual achievements of humanity. Within its walls, a scholar is free to luxuriate in the full range of natural objects and the books that have been written about them. With Bettina or Edith, I came once a week, identifying specimens, measuring shells or crabs, recording geographical data, mining the libraries, and consulting with curators in almost every department of the museum. At Davis, I have become a frequent user of the Museum of Paleontology at the University of California, Berkeley, and of the California Academy of Science in San Francisco.

It is no exaggeration to say that I am in love with museums. In their back rooms, shielded from public view and appreciation, is housed a record of the past and present diversity of life, tended to and studied by a cast of curators and support staff. There are literally millions of specimens—bones, shells, fossils, insects, minerals, dried plants, skins, and human artifacts—all labeled, catalogued, classified, and maintained. Without such collections, we could not identify pests, disease vectors, food species, medically important plants, or the species that suddenly turn up as invaders. We would be unable to trace changes in the geographical distribution of species on time scales ranging from decades to millions of years. Dates of extinction and invasion would remain unknown, and identification of species used in important earlier studies could never be verified.

In the early 1980s, as part of my budding interest in biological invasions, I became curious about the history of the common periwinkle *(Littorina littorea)* in North America and mounted a large historical study of it in museum collections on both sides of the Atlantic. This snail is one of the most abundant inhabitants of rocky shores from Newfoundland to New Jersey, where it thrives on a diet of small seaweeds. It is hard to imagine these shores without the common periwinkle, yet this species was unknown here before the early nineteenth century, when it was found only in its native Europe. The first authenticated report of the common periwinkle in eastern North America is from around 1840 near Halifax, Nova Scotia. Once established, the species spread very rapidly northward and southward, reaching northeastern Maine by 1869 and Atlantic City, New Jersey, by 1892. This spread, which museum holdings document beautifully, was only part of an eventful history of the species in North America.

One of the common periwinkle's predators is the green crab or shore crab, *Carcinus maenas,* another invader from Europe. Like the periwinkle, it was introduced through human agency to eastern North America, but unlike the snail, it was already established before 1820. During most of the nineteenth century, the green crab remained confined to a relatively short segment of coastline between Cape Cod, Massachusetts and New Jersey. After 1900, however, the species began to spread northward, reaching the Atlantic coast of Nova Scotia by 1954.

Here was the perfect opportunity to ask questions about evolution on short time scales. Given that green crabs prey on periwinkles by breaking the shell, predictions can be made about the effects of their northward spread on periwinkles. Many attacks by green crabs are unsuccessful. As in the shells Edith and I studied in the tropics, the repaired breakage of the prey shell resulting from such failed attacks is recorded as a jagged scar cutting across growth increments on the shell surface. If green crabs were important predators of periwinkles, I reasoned, I should observe an increase in the frequency of such scars after the arrival of green crabs north of Cape Cod. Moreover, shell thickness (a measure of the snail's mechanical resistance to breakage) might also be expected to increase. No changes would be expected in Europe, where periwinkles and green crabs have coexisted for a million years or more.

Early naturalists fortunately made large collections of periwinkles in both Europe and North America. Their efforts made it possible for me to measure the incidence of repaired breaks over time among samples from throughout the large geographical range of the species. Edith, Bettina, and I traveled to museums in Philadelphia, New Haven, Cambridge, Amsterdam, Leiden, Brussels, and Copenhagen to examine hundreds of samples collected during the nineteenth and twentieth centuries. Dozens more samples were conveniently located at the Smithsonian, as well as in my own collection.

Satisfyingly, there had been an increase in the frequency of scars in periwinkles after the invasion of green crabs. This rise was not, however, accompanied by an evolutionary response of increased shell thickness. Most populations in Europe showed no change either in the incidence of shell repair or in shell thickness, but Danish ones chronicled a reduction in both. I have no explanation for the Danish data, but it would be interesting to see if ecological or architectural changes involving other species await to be discovered in the museum drawers in Copenhagen. I published the results of this study in *Evolution* in 1982.

Why didn't the periwinkles in North America respond evolutionarily to green crabs? One possibility is that there was insufficient time. Another is that periwinkles have larvae that disperse far from their parents, so that local adaptations are easily swamped by the genes of individuals dispersing from different habitats or regions nearby. A way to test the latter explanation would be to chronicle the history of a snail whose young stages disperse less readily or on a smaller scale.

The dogwhelk, *Nucella lapillus,* was the perfect candidate. Like the periwinkle, this snail occurs on both sides of the Atlantic, but unlike that invader, the dogwhelk has lived in North America for perhaps hundreds of thousands of years.

Again, a rise in the frequency of scars could be documented following the arrival of green crabs north of Cape Cod, but this time the rise was accompanied by the predicted increase in the thickness of the shells' outer lips. Whether this response was in fact evolutionary, caused by natural selection among heritable traits, remains uncertain. Recent experiments by Rich Palmer on dogwhelks indicate that the mere presence of crabs may induce changes in prey shell shape and thickness. Still, these historical patterns could never have been documented without museum samples of common species.

Even if the value of such materials is self-evident to many museum insiders, it remains largely unappreciated by the public. When, in 1994, I wrote an account of the green crab's spread and its consequences for New England snails, I highlighted in a few sentences the pivotal role of museum collections in the reconstruction of this history. Word-counting editors at *Natural History,* the widely read publication of the American Museum of Natural History, snipped out the passage. I protested, pleading that opportunities to bring the work of museums and the value of collections to a broad audience should not be lost. In the end, I prevailed, but the episode graphically exposed the chasm that exists between the research function of the museum and the public's conception of it.

Museums need to educate potential supporters about the value of their work and their materials. They must feature the products of curatorial work in exhibits and even in the museum shop. Exhibits illustrating the evolution or ecology of a group of animals might feature not only specimens from the museum's collections, as is customary, but also the names and works of the scientists who made the discoveries. Taxonomic works or guides by curators may not appeal to many people, but they are unlikely to be bought by those who do not know they exist. Natural history museums the world over are starving for funds at a time when their role as keepers of our biological heritage is more important than ever. With the emphasis in museum exhibits shifting more and more toward entertainment, the gulf between the research and educational departments is widening, and many administrators are responding myopically by cutting back on the study and upkeep of collections. At the museum as at the university, scholarship and teaching enrich each other, and they must be treated as interdependent parts of an indivisible whole.

We build great intellectual edifices in science, and it is fun being an architect of even a very small portion; but buildings rest on foundations, and in natural history those foundations comprise specimens and field observations. A deeply satisfying dimension of the science of natural history is that anyone with curiosity and a love of nature can contribute something of value. We still know almost nothing about the diets, enemies, habitats, and other characteristics of most species, including very common ones. It is with this in mind that I wrote *A Natural History of Shells,* published in 1993 by Princeton University Press. I wanted to transform idle curiosity into a more directed form of questioning, and shells provided an especially esthetic means to that end.

None of this observation would have been possible without freedom and opportunity. I have not been told by universities or government agencies what to study and what not to study, nor has there been censorship of my writing or teaching. Generous funds from the National Science Foundation have enabled me to visit museums, libraries, and field sites throughout the world. When the John D. and Catherine T. MacArthur Foundation completely surprised me with a five-year fellowship in 1992, I experienced another great opportunity to explore new places—Polynesia, New Zealand, Mexico—and other disciplines such as economics. Such precious support gave me the confidence to try new things instead of only continuing to follow familiar paths.

In a world in which habitat destruction and exploitation are proceeding unchecked and outlays for education and science are declining in favor of weapons and entertainment, opportunities for natural historians and many other academics are shrinking. We will all lose much of the beauty of the world and the elegance and power of natural history to enrich our lives as these opportunities continue to erode away.

With opportunity comes responsibility. Whether it was a Princeton fellowship, a grant from the National Science Foundation, or the MacArthur award, I felt the strong urge to meet expectations, to contribute something of value, not to let others down. Most important of these responsibilities are to preserve the freedom to inquire, to ensure that the environments in which we can observe and cherish the diversity of life are preserved, and to serve as curators of our accumulated heritage in the wild as well as in our academic institutions.

Why are northern shells chalky and tropical ones so beautifully crafted and so finely textured? Why do some predatory snails have lip spines while others do not? Why did acorn barnacles replace other kinds of shore animals when they appeared some fifty million years ago? I do not know, but I shall ask. Nature will surprise us if we let her.

GLOSSARY

ADAPTATION   A feature or attribute that confers a survival or reproductive advantage on the organism that bears it, and evolves as the result of a process of natural selection. The word also refers to the evolutionary process that modifies a trait or attribute in a beneficial direction.

AMINO ACID   The chemical building block of a protein molecule. Proteins are formed as long chains of amino acids, of which twenty kinds are commonly encountered in nature.

APERTURE   The opening of a snail's shell, through which the body emerges for feeding and locomotion.

APEX   The tip of a snail's shell. The apex is the oldest part of the shell, close to the point of origin of the spiral coils. The equivalent of the apex in a clam shell is known as the umbo.

BIOGEOGRAPHY   The study of the geographical distribution of organisms.

BIVALVE   An animal whose shell is composed of two movable parts, or valves, usually joined at a flexible hinge. Molluscs with a bivalved shell are commonly called clams, but also include mussels, oysters, and scallops. Bivalved shells have evolved independently in several groups of animals besides molluscs, including brachiopods, crustaceans, and even turtles.

BRACHIOPOD A type of animal with a clam-like, bivalved shell. Although brachiopods look like clams, they are not molluscs at all, but instead comprise a large group unto itself, which was particularly rich in species during the Paleozoic era.

CALCIUM CARBONATE A mineral composed of calcium, carbon, and oxygen, of which many kinds of skeleton are made, including the shells of molluscs, barnacles, and brachiopods, and the stony skeletons of corals and coralline algae. In rock form, calcium carbonate forms limestone.

CAMBRIAN PERIOD The first period of the Paleozoic era, from about 540 to 505 million years ago. At or slightly before the beginning of the Cambrian period, many kinds of animals appear for the first time, including those with mineral skeletons.

CARBONIFEROUS PERIOD The Coal Age, a period during the Late Paleozoic era from about 325 to 275 million years ago. It is best known for its large forests of spore-bearing trees that fossilized to form the great coal deposits of North America and Europe.

CENOZOIC ERA The most recent era in Earth history, from 65 million years ago to the present. This era has been called the age of mammals, but could just as easily be labeled the age of fishes, birds, molluscs, barnacles, and ants, all creatures that proliferated dramatically during this interval.

CHITONS Molluscs whose shell consists of eight curved plates arranged front to back on the upper surface of the animal and connected by a flexible band, or girdle. When dislodged from a rock, a chiton can roll up into a ball, and then flatten out again to cling to another rock. Most chitons feed on seaweeds or small animals, and some excavate deep holes in limestone by gouging the rock with iron-tipped teeth on a flexible tongue ribbon.

CLADE An evolutionary branch, consisting of an ancestor and all its evolutionary descendants.

COEVOLUTION A kind of evolution in which two parties respond to each other. A change in one is followed by a change in the other, and vice versa. Some biologists believe that many insects coevolve with their host food plants. Other possible examples include parasites and their hosts, and such mutually beneficial associations as plants and root fungi, ants and the mushrooms they garden underground, and termites and wood-digesting microorganisms in their digestive systems.

CORAL An animal that typically forms large colonies of interconnected individuals, or polyps. The calcium-carbonate skeletons of corals accumulate to form wave-resistant reefs, which are home to thousands of species of marine organisms. Most reef-building corals harbor single-celled organisms in their tissues that photosynthesize, or make food from inorganic nutrients. Although corals are animals related to sea anemones and jellyfishes, they are ecologically plants, because they produce more oxygen than they consume.

CORALLINE ALGAE Seaweeds that form a hard, stony skeleton.

CRETACEOUS PERIOD The final period of the Mesozoic era, from about 135 to 65 million years ago. The Cretaceous was marked by the rapid and spectacular proliferation of flowering plants and social insects on land, as well as by a rising diversity of life in the sea. The end of the period was marked by a mass extinction that eliminated the dinosaurs as well as vast numbers of planktonic marine organisms.

DEVONIAN PERIOD The interval of the Middle Paleozoic era,

from about 395 to 325 million years ago. This period was marked by the first appearance of forests on land and by the proliferation of tooth-bearing predaceous fishes in the sea. A mass extinction occurred near the end of the period.

DIVERSITY   Variety; in biology, the number of species.

ECOLOGY   The study of the distribution, abundance, and interactions of organisms.

ENDEMIC   Occurring only in one particular geographical region.

ESCALATION   A kind of evolution in which species respond to changes in their enemies or allies.

EVOLUTION   A change in some heritable, or genetic, characteristic of an organism.

FORAMINIFERS   Single-celled protozoas usually with an elaborate skeleton of calcium carbonate.

GASTROPOD   A mollusc that usually has a shell consisting of a single piece, or univalve. Commonly known as snails, gastropods are found on land, in fresh water, and in the sea. Snails that have no shell are usually called slugs.

GENUS   A unit of classification above the species level. One or more species comprise a genus, and one genus or several genera can comprise a family.

GUANO   The accumulated droppings of sea birds. In Peru and elsewhere, guano is mined for phosphate, which is widely used as a fertilizer and in many other applications.

INTERTIDAL ZONE   The part of a seashore that lies between the high and low tide marks. Organisms living in the intertidal zone are alternately immersed in sea water and exposed to air.

JURASSIC PERIOD   The middle period of the Mesozoic era, from about 200 to 135 million years ago. The Late Jurassic was the age of the very largest dinosaurs.

LIMPET   A snail with a cap-shaped shell. Limpets have evolved independently in a large number of snail groups. The shell fits as a cap over the soft body.

MANGROVE   A forest or tree growing in salt water. Mangroves grow on tropical and subtropical shores, where they take the place of salt marshes.

MASS EXTINCTION   A very short interval of time during which a large fraction of species becomes extinct. At least five mass extinctions have been recognized for the eon beginning 550 million years ago.

MESOZOIC ERA   The interval of time from about 250 to 65 million years ago, commonly known as the Age of Dinosaurs.

NATURAL SELECTION   The evolutionary process by which one gene, or unit of heredity, is favored over another because it confers a greater benefit in survival or reproduction. Natural selection is responsible for the evolution of genetic adaptations and also preserves such adaptations once they have evolved.

NERITE   A kind of snail found between tide marks or in the splash zone high on the shore on tropical and subtropical coasts. The West Indian bleeding tooth is a familiar Caribbean example.

PALEONTOLOGY   The study of fossils.

PALEOZOIC ERA   The interval of time from about 540 to 250 million years ago.

PERIOSTRACUM   A horny covering secreted as the outermost layer of mollusc shells.

PERIWINKLE   A kind of snail usually found between tide marks or in the splash zone high on sea shores around the world.

PERMIAN PERIOD   The last period of the Paleozoic era, from about 275 to 250 million years ago. The end of the

Permian was marked by the greatest of all the known mass extinctions, during which perhaps 60 to 70 percent of species became extinct.

PLANKTON   Organisms or communities of organisms floating or swimming in water.

PLEISTOCENE EPOCH   The interval of the Cenozoic era from 1.8 million years to about ten thousand years ago, commonly referred to as the Ice Ages.

PLIOCENE EPOCH   The interval of the Cenozoic era from 5 to 1.8 million years ago. Early representatives of our own genus *Homo* appeared in Africa about 4 million years ago, during the Early Pliocene. Early and Middle Pliocene time was warm, but ice sheets in the northern hemisphere advanced dramatically during later intervals of the epoch.

POLDER   Low-lying land in the Netherlands and Belgium, usually lying below sea level. Polders were created when people drained marshes and bogs.

PUNCTUATED EQUILIBRIA   A pattern of evolution characterized by long episodes during which little or no evolution occurs, and short intervals of more rapid change.

SPECIES   A kind of organism. Individuals comprising a species can fertilize, or be fertilized by, each other in nature in order to produce offspring.

SPIRE   The conical, coiled portion of a snail's shell.

SYMBIOSIS   Living together, usually as a guest organism living on or in a host.

TAXONOMY   The study of the classification of organisms.

TECTONICS   Geological forces that result in continental drift, earthquakes, mountain-building, and other large changes in the Earth's crust.

TRIASSIC PERIOD   The first period the Mesozoic era, from about 250 to 200 million years ago, characterized by the

first appearance of mammals and dinosaurs. A mass extinction occurred near the end of the period.

TRILOBITE   Marine jointed-legged animals with an outer skeleton of calcium carbonate. Trilobites were common during
the Paleozoic era.

UMBILICUS   A slit-like or funnel-shaped cavity in the base of a
snail shell, formed as a space around which the whorls
coil. It may be deep, shallow, or covered in shell callus.

INDEX